Dedication

To my parents, Zhuixian and Yugen,
my wife, Xiaojie, and my daughters,
Alice, Angela, Amy, and Athena.

Disclaimer

Construction Documents and Service (CDS) ARE Mock Exam (Architect Registration Exam) provides general information about Architect Registration Exam. The book is sold with the understanding that neither the publisher nor the authors are providing legal, accounting, or other professional services. If legal, accounting, or other professional services are required, seek the assistance of a competent professional firm.

The purpose of this publication is not to reprint the content of all other available texts on the subject. You are urged to read other materials and tailor them to fit your needs.

Great effort has been taken to make this resource as complete and accurate as possible. However, nobody is perfect and there may be typographical errors or other mistakes present. You should use this book as a general guide and not as the ultimate source on this subject. If you find any potential errors, please send an e-mail to:
info@ArchiteG.com

Construction Documents and Service (CDS) ARE Mock Exam (Architect Registration Exam) is intended to provide general, entertaining, informative, educational, and enlightening content. Neither the publisher nor the authors shall be liable to anyone or any entity for any loss or damages, or alleged loss or damages, caused directly or indirectly by the content of this book.

If you do not wish to be bound by the above, you may return this book to the publisher for a full refund.

Legal Notice

How to Use This Book

We suggest you read *Construction Documents and Service (CDS) ARE Mock Exam (Architect Registration Exam* at least three times:

Read once and cover Chapter One and Two and Appendixes and the related FREE PDF files and other resources. Highlight the information you are not familiar with.

Read twice, focusing on the highlighted information to memorize. You can repeat this process as many times as you want until you master the content of the book. Pay special attention to the following AIA documents: read them at least three times. You may have **many** real ARE CDS division questions based on the following AIA documents listed in **bold** font:

A101–2007, Standard Form of Agreement Between Owner and Contractor where the basis of payment is a Stipulated Sum (CCDC Document 2)
A201–2007, General Conditions of the Contract for Construction
B101–2007 (Former B141–1997), Standard Form of Agreement Between Owner and Architect (RAIC Document 6)

After reviewing these materials, you can do the mock exam, and then check your answers against the answers and explanations in the back, including explanations for the questions you answer correctly. You may have answered some questions correctly for the wrong reason. Highlight the information with which you are not familiar.

Like the real exam, the mock exam includes all three types of questions: Select the correct answer, check all that apply, and fill in the blank.

Review your highlighted information, and do the mock exam again. Try to answer 100% of the questions correctly this time. Repeat the process until you can answer all the questions correctly.

Do the mock exam about one week before the real exam, but at least 3 days before the real exam. You should NOT wait until the night before the real exam and then do the mock exam: If you do not do well, you will go into panic mode and you will NOT have enough time to review your weakness.

Read for the third time the night before the real exam. Review ONLY the information you highlighted, especially the questions you did not answer correctly when you did the mock exam for the first time.

One important tip for passing the graphic vignette section of the ARE CDS division is to become VERY familiar with the commands of the NCARB software. Many people fail the exam simply because they are NOT familiar with the NCARB software and cannot finish the graphic vignette section within the time limit of the exam.

For the two building section vignettes, we include step-by-step solutions using NCARB Practice Program software with many screen-shots so that you can use the book to become familiar with the commands of the NCARB software, even when you do NOT have a computer in front of you. This book is very light and you can carry it around easily. These two features will allow you to review the graphic vignette section whenever you have a few minutes.

All commands are described in an **abbreviated manner**. For example, **Draw > Finished Ceiling** means go to the menu on the left hand side of your computer screen, click **Draw,** and then click **Ceiling** to draw the ceiling. This is typical for ALL commands throughout the book.

The Table of Contents is very detailed so you can locate information quickly. If you are on a tight schedule, you can forgo reading the book linearly and jump around to the sections you need.

All our books, including "ARE Mock Exams Series" and "LEED Exam Guides Series," are available at
GreenExamEducation.com

Check out FREE tips and info at **GeeForum.com**, you can post your questions or vignettes for other users' review and responses.

Table of Contents

Chapter One   **Overview of Architect Registration Exam (ARE)**

   1. Important links to the FREE and official NCARB documents
   2. A detailed list and brief description of the FREE PDF files that you can download from NCARB
      - ARE Guidelines
      - NCARB Education Guidelines
      - Intern Development Program Guidelines
      - The IDP Supervisor Guidelines
      - Handbook for Interns and Architects
      - Official exam guide, <u>references index</u>, and practice program (NCARB software) for each ARE division
      - The Burning Question: Why Do We Need ARE Anyway?
      - Defining Your Moral Compass
      - Rules of Conduct

   1. What is IDP?
   2. Who qualify as interns?

   1. How to qualify for the ARE?
   2. How to qualify for the architect license?
   3. What is the purpose of ARE?
   4. What is NCARB's rolling clock?
   5. How to register for an ARE exam?

6. How early do I need to arrive at the test center?

7. Exam Format & Time
   - Programming, Planning & Practice
   - Site Planning & Design
   - Building Design & Construction Systems
   - Schematic Design
   - Structural Systems
   - Building Systems
   - Construction Documents and Service
8. How are ARE scores reported?
9. Are there a fixed percentage of candidates who pass the ARE exams?
10. When can I retake a failed ARE division?
11. How much time do I need to prepare for each ARE division?
12. Which ARE division should I start first?
13. ARE exam prep and test-taking tips
14. English system (English or inch-pound units) vs. metric system (SI units)
15. Codes and standards used in this book

Chapter Two   Construction Documents & Services (CDS) Division

Chapter One

Overview of Architect Registration Exam (ARE)

A. First Thing First: Go to the website of your architect registration board and read all the requirements for obtaining an architect license in your jurisdiction.
See following link:
http://www.ncarb.org/Getting-an-Initial-License/Registration-Board-Requirements.aspx

B. Download and review the latest ARE documents from NCARB website

1. Important links to the FREE and official NCARB documents
The current version of Architect Registration Exam includes seven divisions:

- Programming, Planning & Practice
- Site Planning & Design
- Building Design & Construction Systems
- Schematic Design
- Structural Systems
- Building Systems
- Construction Documents and Service

Note: Starting July 2010, 2007 AIA Documents apply to Construction Documents & Services Division.

Six ARE divisions have a multiple-choice section and a graphic vignette section. Schematic Design division has NO multiple-choice section, but two graphic vignette sections.

For the vignette section, you need to create the following graphic vignette(s) based on the ARE division you are taking:

Programming, Planning & Practice
Site Zoning

Site Planning & Design
Site Grading
Site Design

Building Design & Construction Systems
Accessibility/Ramp
Stair Design
Roof Plan

Schematic Design
Interior Layout
Building Layout

Structural Systems
Structural Layout

Building Systems
Mechanical & Electrical Plan

Construction Documents & Services
Building Section

There is a tremendous amount of valuable information covering every step of becoming an architect available free of charge at NCARB website: http://www.ncarb.org/

For example, you can find the education guide regarding professional architectural degree programs accredited by the National Architectural Accrediting Board (NAAB), NCARB's Intern Development Program (IDP) guides, initial license, certification and reciprocity, continuing education, etc. These documents explain how you can become qualified to take the Architect Registration Exam.

I find the official ARE Guidelines, exam guide and practice program for each of the ARE divisions extremely valuable. See following link: http://www.ncarb.org/ARE/Preparing-for-the-ARE.aspx

You should definitely start by studying the official exam guide and practice program for the ARE division you are taking.

2. **A detailed list and brief description of the FREE PDF files that you can download from NCARB**
   The following is a detailed list of the FREE PDF files that you can download from NCARB. We list them in an order based on their importance:

   - **ARE Guidelines**: Very important, includes extremely valuable information on the ARE overview, six steps to complete ARE, multiple-choice section, graphic vignette section, exam format, scheduling, sample exam computer screens, links to other FREE NCARB PDF files, practice software for graphic vignettes, etc. You need to read this document at least twice.

- **NCARB Education Guidelines** (Skimming through it should be adequate)

- **Intern Development Program Guidelines**: Important information on IDP overview, IDP steps, IDP reporting, IDP basics, work settings, training requirements, supplementary education (core), supplementary education (elective), core competences, next steps, and appendices. Most of NCARB's 54-member boards have adopted the IDP as a prerequisite for initial architect licensure. That is the reason you should care about it. IDP costs $350 for the first three years, and then $75 annually. The fees are subject to change, and you need to check the NCARB website for the latest information. You need to report your IDP experience <u>no longer than every six months</u> and within two months of completion of each reporting period (the **Six-Month Rule**). You need to read this document <u>at least twice</u>. It has a lot of valuable information.

- **The IDP Supervisor Guidelines** (Skimming through it should be adequate. You should also forward a copy of this PDF file to your IDP supervisor.)

- **Handbook for Interns and Architects** (Skimming through it should be adequate)

- **Official exam guide, <u>references index</u>, and practice program (NCARB software) for each ARE division**
  Specific information for ARE divisions (You just need to focus on the documents related to the ARE divisions you are currently taking and read them at least twice. Make sure you install the practice program and become very familiar with it. The real exam is VERY similar to the practice program):

  a. **Programming, Planning & Practice (PPP)**: Official exam guide and practice program for PPP division
  b. **Site Planning & Design (SPD)**: Official exam guide and practice program (computer software) for SPD division
  c. **Building Design & Construction Systems (BDCS)**: Official exam guide and practice program for BDCS division
  d. **Schematic Design (SD)**: Official exam guide and practice program for SD division
  e. **Structural Systems (SS)**: Official exam guide, <u>references index</u>, and practice program for SS division
  f. **Building Systems (BS)**: Official exam guide and practice program for BS division
  g. **Construction Documents and Service (CDS)**: Official exam guide and practice program for CDS division

- **The Burning Question: Why Do We Need ARE Anyway?** (Skimming through it should be adequate)

- **Defining Your Moral Compass** (Skimming through it should be adequate)

- **Rules of Conduct** (Skimming through it should be adequate). Available as a FREE PDF file at:
  http://www.ncarb.org/

C. The Intern Development Program (IDP)

1. What is IDP?

It is a comprehensive training program jointly developed by the National Council of Architectural Registration Boards (NCARB) and the American Institute of Architects (AIA) to ensure the interns obtain the necessary skills and knowledge to practice architecture <u>independently</u>.

2. Who qualify as interns?

Per NCARB, the following individuals qualify as interns:

a. Graduates from NAAB-accredited programs

b. Architecture students who acquire acceptable training prior to graduation

c. Other qualified individuals identified by a registration board

D. Overview of Architect Registration Exam (ARE)

1. How to qualify for the ARE?

A candidate needs to qualify for the ARE via one of NCARB's member registration boards, or one of the Canadian provincial architectural associations.

Check with your Board of Architecture for specific requirements.

For example, in California, a candidate must provide verification of a minimum of <u>five</u> years of education and/or architectural work experience to qualify for the ARE.

Candidates can satisfy the five-year requirement in a variety of ways:

- Provide verification of a professional degree in architecture through a program that is accredited by NAAB or CACB.

  OR

- Provide verification of at least five years of educational equivalents.

  OR

- Provide proof of work experience under the direct supervision of a licensed architect

2. **How to qualify for the architect license?**

Again, each jurisdiction has its own requirements. It typically requires a combination of about <u>eight</u> years of education and experience, as wells as passing ARE exams. See following link:

http://www.ncarb.org/Reg-Board-Requirements

For example, to become a licensed architect in California, you need:

- Eight years of post-secondary education and/or work experience as evaluated by the Board (including at least one year of work experience under the direct supervision of an architect licensed in a U.S. jurisdiction or two years of work experience under the direct supervision of an architect registered in a Canadian province)
- Completion of the Comprehensive Intern Development Program (CIDP) and the Intern Development Program (IDP)
- Successful completion of the Architect Registration Examination (ARE)
- Successful completion of the California Supplemental Examination (CSE)

California does NOT require an accredited degree in architecture for examination and licensure. However, many other states require an accredited degree for licensure.

3. **What is the purpose of ARE?**

The purpose of ARE is NOT to test a candidate's competency on every aspect of architectural practice. Its purpose is to test a candidate's competency on providing professional services to protect the <u>health, safety, and welfare</u> of the public. It tests candidates on the <u>fundamental</u> knowledge of pre-design, site design, building design, building systems, and construction documents and services.

ARE tests a candidate's competency as a "specialist" on architectural subjects. It also tests her abilities as a "generalist" to coordinate other consultants' works.

You can download the exam content and references for each of the ARE divisions at the following link:

http://www.ncarb.org/are/40/StudyAids.html

4. **What is NCARB's rolling clock?**

a. Starting on January 1, 2006, a candidate MUST pass ALL ARE sections within 5 years. A passing score for an ARE division is only valid for 5 years, and a candidate has to retake this division if he has NOT passed all divisions within the 5-year period.

b. Starting on January 1, 2011, a candidate who is authorized to take ARE exams MUST take at least one division of the ARE exams within 5 years of the authorization. Otherwise, the candidate MUST apply for the authorization to take ARE exams from an NCARB member board again.

These rules are created by the **NCARB's rolling clock** resolution passed by NCARB council in 2004 NCARB Annual Meeting.

5. **How to register for an ARE exam?**
   Go to the following website and register:
   http://www.prometric.com/NCARB/default.htm

6. **How early do I need to arrive at the test center?**
   At least 30 minutes BEFORE your scheduled test time, OR you may lose your exam fee.

7. **Exam Format & Time**
   All ARE divisions are administered and graded by computer. Their detailed exam format
   and testing time are as follows:

   1) **Programming, Planning & Practice (PPP)**
      The **Programming, Planning & Practice** division of the Architect Registration Exam
      (ARE) includes a multiple-choice (MC) section and a graphic vignette section, and lasts
      a total of 4 hours. It includes:

Introduction Time:	15 minutes	
MC Testing Time:	**2 hours**	**85 items**
Scheduled Break:	15 minutes	
Introduction Time:	15 minutes	
Graphic Testing Time:	**1 hour**	**Site Zoning (1 vignette)**
Exit Questionnaire:	15 minutes	
Total Time	**4 hours**	

   2) **Site Planning & Design (SPD)**

Introduction Time:	15 minutes	
MC Testing Time:	**1.5 hours**	**65 items**
Scheduled Break:	15 minutes	
Introduction Time:	15 minutes	
2 Graphic Vignettes:	**2 hours**	**Site Grading, Site Design**
Exit Questionnaire:	15 minutes	
Total Time	**4.5 hours**	

   3) **Building Design & Construction Systems (BDCS)**

Introduction Time:	15 minutes	
MC Testing Time:	**1.75 hours**	**85 items**
Scheduled Break:	15 minutes	
Introduction Time:	15 minutes	
3 Graphic Vignettes:	**2.75 hours**	**Accessibility/Ramp, Stair Design, Roof Plan**
Exit Questionnaire:	15 minutes	
Total Time	**5.5 hours**	

4) Schematic Design (SD)

Introduction Time:	15 minutes	
Graphic Testing Time:	**1 hour**	**Interior Layout (1 vignette)**
Scheduled Break:	15 minutes	
Introduction Time:	15 minutes	
Graphic Testing Time:	**4 hours**	**Building Layout (1 vignette)**
Exit Questionnaire:	15 minutes	
Total Time	**6 hours**	

5) Structural Systems (SS)

Introduction Time:	15 minutes	
MC Testing Time:	**3.5 hours**	**125 items**
Scheduled Break:	15 minutes	
Introduction Time:	15 minutes	
Graphic Testing Time:	**1 hour**	**Structural Layout (1 vignette)**
Exit Questionnaire:	15 minutes	
Total Time	**5.5 hours**	

6) Building Systems (BS)

Introduction Time:	15 minutes	
MC Testing Time:	**2 hours**	**95 items**
Scheduled Break:	15 minutes	
Introduction Time:	15 minutes	
Graphic Testing Time:	**1 hour**	**Mechanical & Electrical Plan (1 vignette)**
Exit Questionnaire:	15 minutes	
Total Time	**4 hours**	

7) Construction Documents and Service (CDS)

Introduction Time:	15 minutes	
MC Testing Time:	**2 hours**	**100 items**
Scheduled Break:	15 minutes	
Introduction Time:	15 minutes	
Graphic Testing Time:	**1 hour**	**Building Section (1 vignette)**
Exit Questionnaire:	15 minutes	
Total Time	**4 hours**	

8. How are ARE scores reported?

All ARE scores are reported as Pass or Fail. ARE scores are processed within 4 to 6 weeks, and then sent to your Board of Architecture. Your board then does additional processing and forwards the scores to you.

9. Are there a fixed percentage of candidates who pass the ARE exams?
No, there is NOT a fixed passing or failing percentage. If you meet the minimum competency required to practice as an architect, you pass. The passing scores are the same for all Boards of Architecture.

10. When can I retake a failed ARE division?
You can only take the same ARE division once within a 6-month period.

11. How much time do I need to prepare for each ARE division?
You need about 40 hours to prepare for each ARE division. You need to set a realistic study schedule and stick with it. Make sure you allow time for personal and recreational commitments. If you are working full time, my suggestion is that you allow no less than 2 weeks but NOT more than 2 months to prepare for each ARE division. You should NOT drag out the exam prep process too long and risk losing your momentum.

12. Which ARE division should I start first?
It is a matter of personal preference, and you should make the final decision.

Some people like to start with the easier divisions and pass them first. This way, they build more confidence as they study and pass each division.

Some people like to start with the more difficult divisions and pass them first. This way, if they fail a difficult, as they study the other divisions, the clock start to clicks, and they can reschedule the failed division six months later.

Programming, Planning & Practice (PPP) and Building Design & Construction Systems (BDCS) divisions often include some content from the Construction Documents and Service (CDS) division. It may be a good idea to start with CDS and then schedule the exams for PPP and BDCS soon after.

13. ARE exam prep and test-taking tips
You can start with Construction Documents and Service (CDS) and Structural Systems (SS) first because both divisions gave a limited scope, but you may want to study building regulations and architectural history (especially famous architects and buildings that set the trends at critical turning points) before you take other divisions.

Do mock exams and practice questions and vignettes, including those provided by NCARB's practice program and this book, to hone your skills.

Form study groups and learn the exam experience of other ARE candidates. The forum at our website is a helpful resource. See the following link:
http://GreenExamEducation.com/

Take the ARE exams as soon as you become eligible, since you probably still remember portions of what you learned in architectural schools, especially structural and architectural history. Do not make excuses for yourself and put off the exams.

The following test-taking tips may help you:
- Pace yourself properly: You should spend about one minute for each Multiple-Choice (MC) question, except for the SS division. You should spend about one and a half minutes for each Multiple-Choice (MC) question for the SS division.
- Read the questions carefully and pay attention to words like *best, could, not, always, never, seldom, may, false, except,* etc.
- For questions that you are not sure of, eliminate the obvious wrong answer and then make an educated guess. If you do NOT answer the question, you automatically lose the point. If you guess, you have a chance to gain the point.
- If you have no idea what the correct answer is and cannot eliminate any obvious wrong answer, then do not waste too much time on the question; just pick a guess answer. The key is, try to use the same guess answer for all of the questions for which you have no ideas at all. For example, if you choose "d" as the guess answer, then you should be consistent and use "d" as the guess answer for all the questions for which you have no ideas at all. That way, you likely have a better chance at guessing more correct answers.
- Mark the difficult questions, answer them, and come back to review them AFTER you finish all MC questions. If you are still not sure, go with your first choice. Your first choice is often the best choice.
- You really need to spend time practicing to become VERY familiar with NCARB's graphic software and know every command well. This is because ARE graphic vignette is a timed test, and you do NOT have time to think about how to use the NCARB's graphic software during the test. Otherwise, you will NOT be able to finish your solution to the vignette on time.
- ARE exams test a candidate's competency on providing professional services to protect the <u>health, safety, and welfare</u> of the public. Do NOT waste your time on aesthetic or other design elements not required by the program.

ARE exams are difficult, but if you study hard and prepare well, combined with your experience, IDP training, and/or college education, you should be able to pass all divisions and eventually be able to call yourself an architect.

14. English system (English or inch-pound units) vs. metric system (SI units)
This book is based on the English system or English units; the equivalent value in metric system or SI units follows in parentheses. All SI dimensions are in millimeters unless noted otherwise. The English or inch-pound units are based on the module used in the U.S. The SI units noted are simple conversions from the English units for information only and are not necessarily according to a metric module.

15. Codes and standards used in this book
We use the following codes and standards used in this book:
American Institute of Architects, Contract Documents, Washington, DC.
Canadian Construction Documents Committee, CCDC Standard Documents, 2006, Ottawa.

Chapter Two

Construction Documents & Services (CDS) Division

A. General Information

1. **Exam Content**
   An architect should be able to prepare contract documents and perform construction administration.

   The exam content for the CDS division of the ARE includes:
   1) Codes & regulations
   2) Environmental issues
   3) Construction drawings & project manual
   4) Project & practice management
      - Cost
      - Scheduling & Coordination
      - Project Delivery (including submittals)
      - Contracts & Legal Issues

   For the graphic vignette, you will be required to draw a building section that incorporates structural, mechanical, and lighting systems as well as life safety concerns.

2. **Official exam guide and practice program for CDS division**

   You need to read the official exam guide for CDS division at least twice. Make sure you install the CDS division practice program and become very familiar with it. The real exam is VERY similar to the practice program.

B. The most important documents/publications for CDS division of the ARE exam
Based on our research, the most important documents/publications for CDS division of the ARE exam are:

1. **American Institute of Architects (AIA) Documents**
   Reading the summary of the AIA Documents is NOT adequate. You need to read the complete text. Fortunately, you do NOT have to read all the AIA documents.

Two possible solutions:

a. Buy ONLY the AIA documents you need from your local AIA office. The following AIA documents are important (especially the AIA documents in **bold** font, read them at least three times. You may have **many** real ARE CDS division questions based on the following AIA documents listed in **bold** font):

- **A101–2007, Standard Form of Agreement Between Owner and Contractor where the basis of payment is a Stipulated Sum (CCDC Document 2)**
- **A201–2007, General Conditions of the Contract for Construction**
- **A503–2007, Guide for Supplementary Conditions**
- **A701–1997, Instructions to Bidders (CCDC Document 23)**
- **B101–2007 (Former B141–1997), Standard Form of Agreement Between Owner and Architect (RAIC Document 6)**

You can find FREE sample forms and commentaries for AIA documents A201 & B101 at the following link:
http://www.aia.org/contractdocs/aiab081438

- **C401–2007 (Former C141–1997), Standard Form of Agreement Between Architect and Consultant**

There are some <u>major changes</u> to the older version of sample C141–1997.

A FREE older version of sample C141–1997 is available at the following link:
https://app.ncarb.org/are/StudyAids/_C141.pdf

- G701–2001, Change Order
- G702–1992, Application and Certificate for Payment
- G704–2000, Certificate of Substantial Completion

AIA updates AIA documents roughly every 10 years. Please note that AIA does NOT update <u>all</u> AIA documents at the same time. For example, A701–1997, Instructions to Bidders (CCDC Document 23) is still the most current form.

See the AIA documents price list at the following link for the latest edition of AIA documents:
http://www.aia.org/aiaucmp/groups/aia/documents/pdf/aias076346.pdf

b. Buy AHPP, and it has a CD that include the sample AIA documents. AHPP itself is also a very important publication for CDS division. See information at following Item 2.

2. ***The Architect's Handbook of Professional Practice* (AHPP)**
Demkin, Joseph A., AIA, Executive Editor. *The Architect's Handbook of Professional Practice* (AHPP). The American Institute of Architects & Wiley, latest edition. This comprehensive book covers all aspect of architectural practice, including two CDs

containing the sample AIA contract documents. You may have a few real ARE CDS division questions based on this publication. You need to read through this book a few times and know some of the basic architectural practice elements, such as firm legal structures, marketing and outreach, the delivery methods and compensation (design-bid-build, construction management, and design-build), contracts and agreements, and managing disputes (mediation, arbitration and litigation).

3. *International Building Code* **(IBC)**
   International Code Council, Inc. (ICC, 2006), *International Building Code* (IBC). You may have a few real ARE CDS division questions based on this publication. You need to become familiar with some of the commonly used code sections, such as allowable areas and allowable areas increase, unlimited areas, egresses, width and numbers of exits required, minimum exit passage width, occupancy groups and related exit occupancy load factors, types of construction, minimum number of required plumbing fixtures required, etc.

   See the following link for some FREE IBC code sections citations:
   http://publiccodes.citation.com/icod/ibc/2006f2/index.htm?bu=IC-P-2006-000001&bu2=IC-P-2006-000019

4. *Architectural Graphic Standards*
   Ramsey, Charles George, and John Ray Hoke Jr. *Architectural Graphic Standards*. The American Institute of Architects & Wiley, latest edition. There may be a few questions asking you what some basic graphic symbols mean. This is a good book to skim through.

5. **Construction Specifications Institute (CSI) MasterFormat &** *Building Construction*
   You need to become familiar with the new 6-digit CSI Construction Specifications Institute (CSI) MasterFormat. You may have a few real ARE CDS division questions based on this publication. You need to know what items/materials belong to which CSI MasterFormat specification section, and remember some major sections names and numbers, such as Division 9 is Finishes, and Division 5 is Metal, etc. My other book, *Building Construction*, has detailed discussions on CSI MasterFormat specification sections.

   See the Appendix of this book for other official reference materials suggested by NCARB.

C. Tips, Solution and Notes on NCARB traps

1. Tips for graphic vignette section

Overall strategy: Draw from bottom to top, start with the first floor, finish it, and then work on the second floor.

Understand **several key items/traps NCARB** wants to test you:

1) Make sure you draw the **grade line** and align your section to the left exterior wall on plan, otherwise you will fail. The left side of your section should be placed along the line that is designated "**exterior face of section**."

2) You still need to draw **slab on grade** after you draw the grade line. Many people miss this step.

3) Pay attention to the depth of frost line, and **bottoms** of all the exterior footings should be set **at** the frost line. The depths of footings for interior bearing walls should be right under the slab.

4) There is typically one large room that has a high ceiling and has deeper joists.

5) The program typically requires you to place **ALL** ducts below the joists. So, do NOT just look at the duct sizes at the section cut line location, you probably need to look for the **deepest** duct size for the **entire** floor that you are working on, and use the **deepest** duct size to set your clearance under the joists. Show the ducts close to the bottom of the joists.

*Note: The double height space is an exception because it has its own joist depth and you just need to use the **deepest** duct size for the double height space to set the clearance under the joists for the double height space.*

6) Do NOT forget the clearance needed for the light fixtures.

7) Draw the joists elevations or sections per floor plan.

8) The **corridor walls are fire-rated**, and it should extend to the deck above.

9) The height of parapet has to meet the program requirements.

2. Step-by-step solution to the official NCARB CDS practice program problem: graphic vignette section

The official exam guide gives a passing solution and a failing solution to the sample graphic vignette problem, but it does NOT show you the step-by-step details, and it is NOT tied to official NCARB CDS practice program.

I am going to fill in the blanks here. I am going to offer you step-by-step instructions, and tie them to the commands of the official NCARB CDS practice program (graphic software).

You really need to spend time practicing to become VERY familiar with NCARB's graphic software. This is because ARE graphic vignette is a timed test, and you do NOT have time to think about how to use the NCARB's graphic software during the test. Otherwise, you will NOT be able to finish your solution to the vignette on time.

The following solution is based on the official NCARB CDS practice program for **ARE 4.0**. Future versions of ARE may have some minor changes, but the principles and fundamental elements should be the same. The official NCARB CDS practice program has very few changes since its introduction and the earlier versions are VERY similar to, or probably exactly the same as, the current ARE 4.0. The actual graphic vignette of the CDS division should be VERY, VERY similar to the practice graphic vignette in the NCARB CDS practice program.

Tips:
1) You need to install the NCARB CDS practice program, and become familiar with it. I am NOT going to repeat the problem description and the first and second floor plans here since they are already available in the NCARB practice program.

   See the following link for a FREE download of the NCARB practice program: http://www.ncarb.org/ARE/Preparing-for-the-ARE.aspx

2) Review the general test directions, vignette directions, program, and tips carefully.
3) Press the space bar to go to the work screen.

Solution:
1) Draw the **grade line**:

   Draw > Grade Line (This is the actual command in NCARB CDS practice program. We shall use this format to note the command for the entire book)

   This is the most important step. Your solution will fail if you miss this step. Make sure you leave enough space to draw both the first and second floor sections. Otherwise, you may have to spend more time moving all the elements that you draw later to make room for the second floor section.

   Note: Some people prefer to draw the grade line near the section cut line, and overlap the section with the floor plans. This way, they can see the relationship between the section and the floor plans right there without zoom in and out. I personally find this approach confusing and make the drawings too hard to read. Of course, you can have your own preference. Ultimately, it is your choice and work habit.

2) Draw the **exterior walls and bearing walls**:
   You can use **sketch > line** to project the exact location of the walls from the plans to the section.

   Draw > Exterior/Bearing Wall

   You can draw the walls higher than their actual heights for now, and adjust the heights later.

   Note: The left side of your section must align with the line noted as "EXTERIOR FACE OF SECTION" on the plans. Otherwise, your solution will automatically fail.

3) Draw the **slab on grade**:

Draw > Slab on Grade (Figure 2.1)

*Note: The slab is NOT continuous, and it stops at the interior bearing wall, i.e., interior bearing wall sits directly on the foundation below and divides the slab into two segments on this section. You can use **Zoom** command to zoom in and adjust or move the element you draw to a more accurate location.*

You should place the <u>bottom</u> of the slab right on top of the grade line.

For example, we notice the slab on grade is too far above the grade line, and we zoom in to adjust it to right above the grade line:

Zoom > click on the rectangle area that you want to zoom in

Use **Move, Adjust** to move the slab on grade to the desired location, and then click on the location to place the slab on grade on the new location. (Figure 2.2)

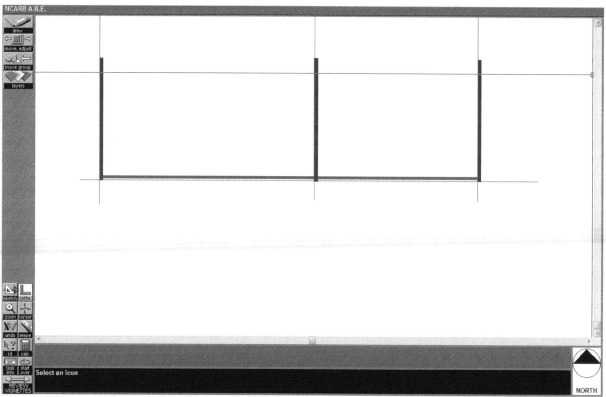

Figure 2.1 Draw Grade Line, Exterior/Bearing Walls, and Slab on Grade.

4) Draw **foundation with footing under the exterior walls**

Zoom in to the area you want to work on

Draw > Foundation with Footing (Figure 2.3)

Per the program of this vignette, the frost depth is 5'-0" (1524, which means 1524 millimeter or mm, typical for the entire book), so make sure the <u>bottom</u> instead of the <u>top</u> of the footing is placed <u>exactly</u> at 5'-0" (1524) below the grade line to match the frost depth. The foundation depth automatically shows on the lower left hand corner of the computer screen when you draw the foundation with footing.

Use **Move, Adjust** to move exterior foundation with footing to align with the exterior wall (Figure 2.4).

Use the same procedure to draw the other exterior foundation with footing.

Note:
1. *You should always place the <u>bottom</u> instead of the <u>top</u> of the exterior footings at frost line. The reasons are:*

   Most soils expand and heave when they freeze.

   *The **frost line** is the depth at which the moisture present in the soil is expected to freeze. Once the bottom of the footing is placed at the frost line, the ground will act as a barrier to insulate the soil below the footing from freezing in the winter, and thus prevent the soils below the frost line from expanding and heaving.*

   If you place the <u>top</u> of the footing at the frost line, it will still work, but it will cost MORE money, and it is not as good as placing the <u>bottom</u> of the footing at the frost line. You are supposed to draw the <u>minimum</u> depth for the footings for the CDS graphic vignette.

2. *You should always align the <u>top</u> of the foundation with footings with the grade line (the <u>bottom</u> of the slab), NOT the <u>top</u> of the slab. The reason is:*

   If you align the <u>top</u> of the foundation with footings with the <u>top</u> of the slab, and you draw a 5'-0" deep foundation with footing using the NCARB graphic software, your bottom of the footing will be a few inches higher than the frost line because of the slab thickness, and your solution will probably fail.

5) Use the same procedure to draw the **footing with foundation for the interior bearing wall**, except the top of the foundation should be placed <u>right below the bottom of the slab</u>, NOT close to the frost depth. The interior bearing wall is typically thicker, and you can confirm it is an interior bearing wall by turning on only the layers for the 1st floor, or only the layer for the 2nd floor to see the related structural joists.

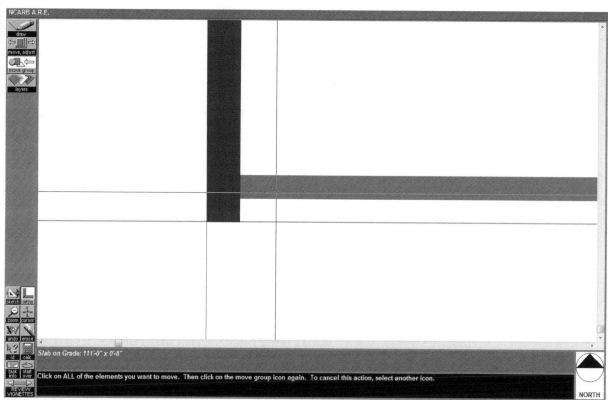

Figure 2.2   Zoom in and use **Move, Adjust** to move the slab on grade to an accurate location.

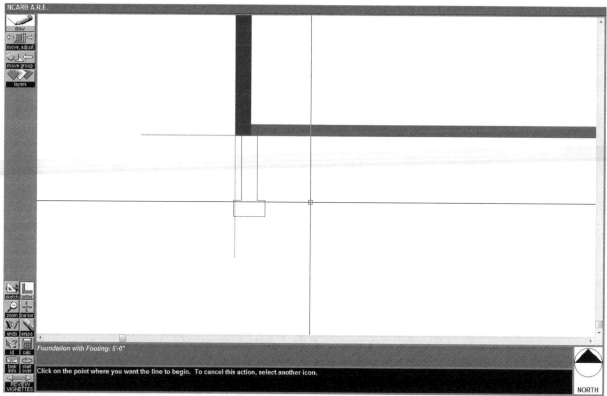

Figure 2.3 Draw exterior foundation with footing.

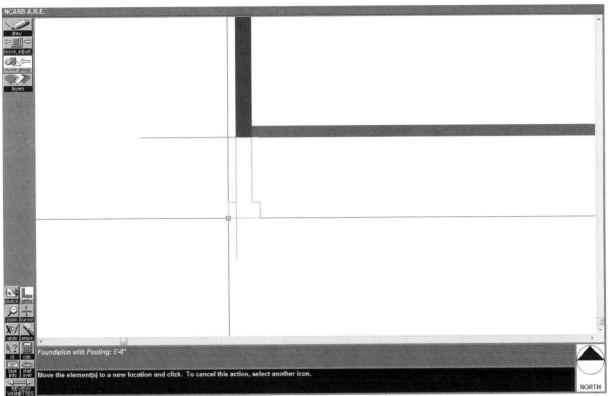

Figure 2.4   Use **Move, Adjust** to move exterior foundation with footing to align with the exterior wall.

Figure 2.5   Use **Sketch > Rectangle** to establish the ceiling height, and use **Draw > Finished Ceiling** to draw the ceiling.

6) Use **Sketch > Rectangle** to draw a 15'-0" (4572) high rectangle to establish the ceiling height for the laboratory.

Note: The ceiling height is measured from the <u>top</u> of the slab on grade to the <u>bottom</u> of the ceiling.

7) Use **Draw > Finished Ceiling** to draw the ceiling for the laboratory (Figure 2.5).
8) Use **Sketch > Rectangle** to draw an 8" (203) high rectangle to establish the **clearance for light fixtures**.
9) The laboratory is a **double height space** and has its own joist depth. The ducts at the section cut line are 8x8 (203x203). However, the deepest duct for the laboratory is 24x20 (609x508) with a depth of 20" (508). Since **ALL** ducts are placed below the joists, you should use 20" (508) as the duct clearance space for the laboratory.
10) Use **Sketch > Rectangle** to draw a 20" (508) high rectangle right above the light fixture clearance line to establish the **clearance for the ducts**.
11) Use **Layers** to turn on only the 2nd floor layers. Use **Sketch > Line** to project the locations of the first and last joists for the laboratory space.
12) For the laboratory space: since the joists are cut by the section cut line, they should be shown as a sectional view. Since the 2nd floor plan notes 32" (812) joists @ 5'-0" (60" or 1524) O.C., we use **Draw > Joists with Deck (Section) > Depth 32" (812) > Spacing 60"** (60" or 1524) to draw the joists for the laboratory space.
13) Use **Move Group** to move the entire joist and align the bottom of the joists with the top of the duct clearance line, and align the 1st joist and last joist with the ones on plan (Figure 2.6).
14) Use **Zoom** and **Move, Adjust** to extend both ends of the deck to 4" into the exterior wall and the interior bearing wall.

*Note: You can use **Move Group** to move the entire joist run, and **Move, Adjust** to adjust the locations of the joist within the joist run. Make sure the locations of your joists on the section MATCH the locations of the joists on the floor plans. Otherwise, you may fail this vignette.*

15) Use **Draw > Duct > Width 8" (203) > Depth 8"** (203) to draw the 8x8 (203x203) duct at section cut line, align them with floor plan, and align the top of the duct to the bottom of the joists (Figure 2.7).
16) Use **Sketch > Rectangle** to draw a 2'-0" (610) high rectangle to establish the top of parapet height.
17) Use **Move, Adjust** to adjust the top of parapet to 2'-0" (610) above the highest adjacent roof, as required by the program (Figure 2.8).
18) Repeat steps 7) to 18) to draw the remaining rooms for the 1st and 2nd floors (Figure 2.9).
19) Use **Sketch > Hide Sketches Elements** to show the completed section without the sketch elements (Figure 2.10).

You should practice aforementioned steps 1) to 19) a few times to become familiar with the process, understand the pitfalls, and master the NCARB CDS practice program. You should be able to finish the graphic vignette within the required 1-hour testing time.

3. **Notes on NCARB traps:**

   Pay attention to the following NCARB traps:
   - In normal practice, most architects place the larger ducts parallel to the joists and **between** the joists to reduce interstitial height. The NCARB program for the CDS graphic vignette does NOT allow this good practice, and **ALL** ducts must be placed **below** the joists.
   - The interstitial height may NOT be determined by the deepest duct plus the deepest joists. It is determined by the largest joist-duct **combination**, one above the other in the **same** location.

     For example, if the largest duct is 32x24 (813x610), and the joist size at the location of the 32x24 (813x610) duct is 24" (610) joists at 2'-0" (610) O.C., then the clearance needed for this space is 24" (610) + 24" (610) = 48" (1220).

     On the other hand, if there is another duct that is 28x20 (711x508), and the joist size at the location of the 28x20 (711x508) duct is 36" (914) joists at 5'-0" (1524) O.C., then the clearance needed for this space is 20" (508) + 36" (914) = 56" (1422). This is the **worst** case for the entire floor, and you should use this **combination** to set the interstitial space.

     If you use the largest joist depth plus the largest duct depth: 36" (914) + 24" (610) = 60" (1524). Your interstitial space is 4" (102) TOO big.

   Now, let us go back to this specific NCARB CDS graphic vignette:

   - **For the remaining rooms of the 1st floor:** since all the joists are the same size, we can use the **deepest** duct to establish the duct clearance for the 1st floor joists. The ducts at the section cut line location are 24x16 (610x406) and 12x8 (305x1626). You need to show the duct sections at the correct sizes and align them to the bottom of the joist. However, the **deepest** duct for the remaining rooms of the 1st floor is 32x24 (813x610) with a depth of 24" (610). Since **ALL** ducts are placed below the joists, you should use 24" (610) as the duct clearance depth for the remaining rooms of the 1st floor.
   - **For the 2nd floor:** since all the joists are the same size, we can use the **deepest** duct to establish the duct clearance for the 2nd floor joists. The duct at the section cut line location is 20x12 (508x305). You need to show the duct section at the correct size and align them to the bottom of the joist. However, the **deepest** duct for the entire 2nd floor is 32x 20 (813x508) with a depth of 20" (508). Since **ALL** ducts are placed below the joists, you should use 20" (508) as the duct clearance depth for the 2nd floor.

     Note: You should always place the ducts right below the bottom of the joist for easier maintenance access and make more room for light fixtures and fire sprinklers below.

- Do NOT forget the 8" (203) clearance height needed for light fixtures.
- The corridor walls are fire-rated, and should be extended to the bottom of deck. Use **Draw > Interior Fire-Rated Partition** to draw them.
- The other interior walls are NOT fire-rated, and should be extended to 6" (152) above the finished ceiling. Use **Draw > Interior Partition** to draw them.
- Align the section elements (ducts, joists, and interior partitions) with those on the floor plans.
- Make sure you draw the ducts and joists at the <u>correct</u> sizes (and <u>correct</u> spacing for joists) per the plans.
- Draw room heights, parapet heights, clearance for ducts, and light fixtures per the **minimum dimensions required** by the program.

4. **A summary of the critical dimensions**
   For your convenience, the following is a summary of the critical dimensions:

Frost depth (<u>Bottom</u> of the exterior footings)
5'-0" (1524) below the grade line (or the <u>bottom</u>, NOT the top of the slab)

Laboratory (Double Height Space)
Ceiling – 15'-0" (4572)
Lights – 8" (203)
Worst case duct depth – 20" (508)
Joists - 32" @ 5'-0" O.C. (813 @ 1524 O.C.)
Deck - 4" (100), automatically drawn on top of the joists
Duct size at section cutline: 8x8 (203 x 203)
Parapet – 24" (610)

First Floor
Ceiling – 8'-4" (2540)
Lights – 8" (203)
Worst case duct depth – 24" (610)
Joists – 24" @ 2'-0" O.C. (610 x 610 O.C.)
Deck - 4" (100), automatically drawn on top of the joists
Duct sizes at section cutline: 24x16 (610 x 406) and 12x8 (305 x 203)

Second Floor
Ceiling – 9'-0" (2743)
Lights – 8" (203)
Worst case duct depth – 20" (508)
Joists – 24" @ 5'-0" O.C. (610 @ 1524 O.C.)
Deck - 4" (100), automatically drawn on top of the joists
Parapet – 24" (610)
Duct size at section cutline: 20x12 (508 x305)

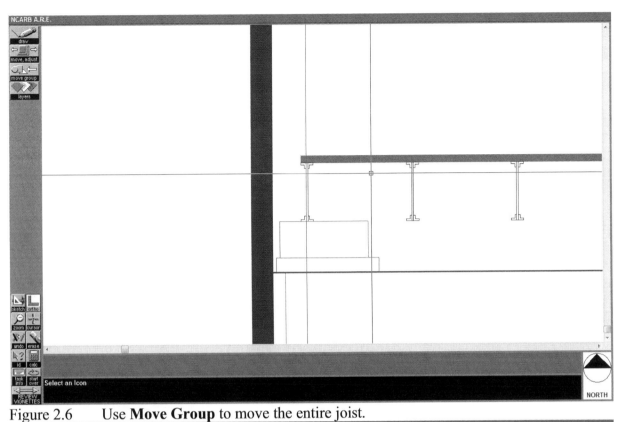

Figure 2.6   Use **Move Group** to move the entire joist.

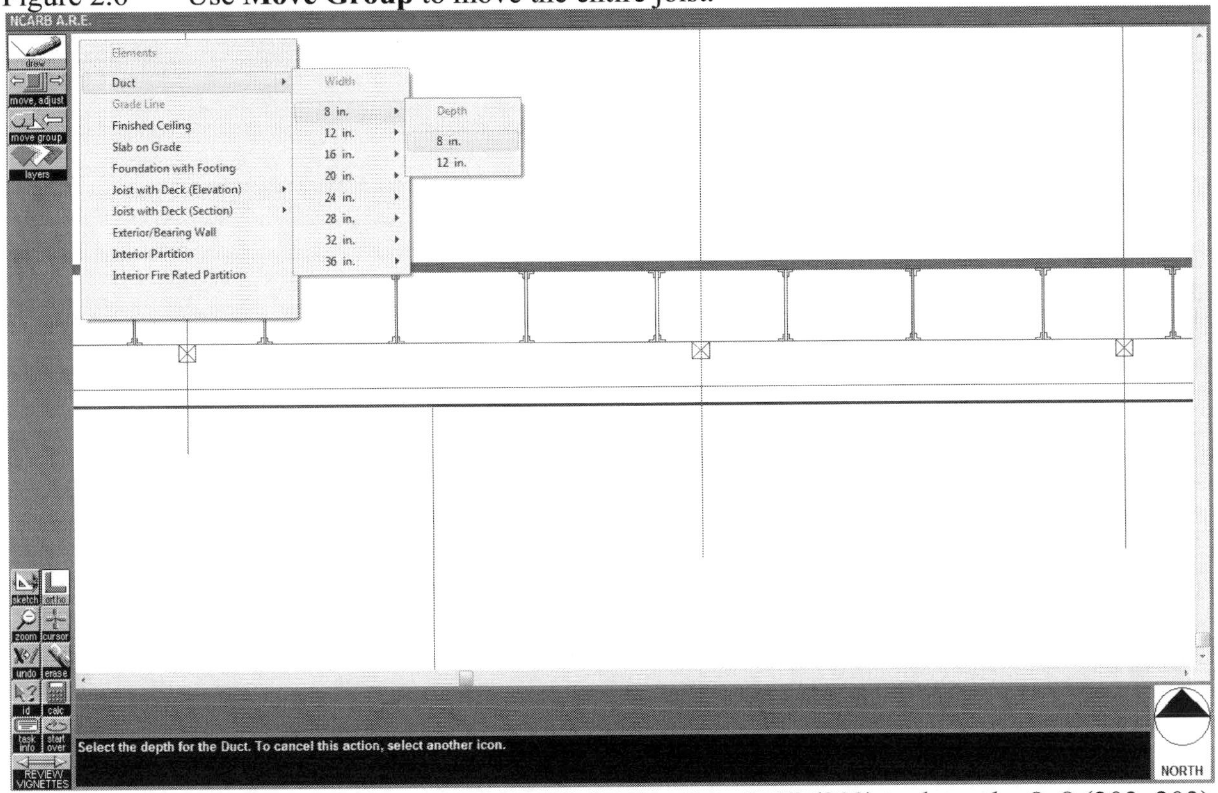

Figure 2.7   Use **Draw > Duct > Width 8"** (203) > **Depth 8"** (203) to draw the 8x8 (203x203) duct at section cut line.

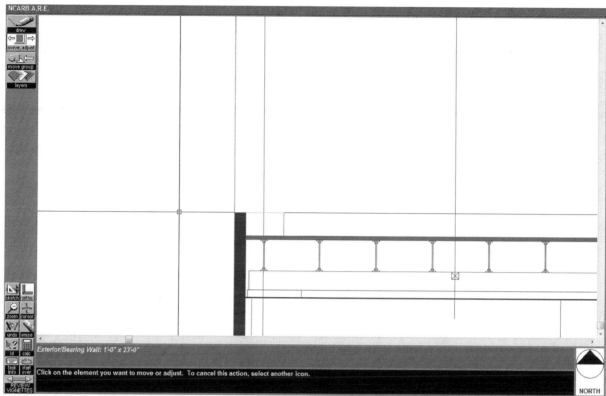

Figure 2.8   Use **Move, Adjust** to adjust the top of parapet to 2'-0" (610) above the highest adjacent roof, as required by the program.

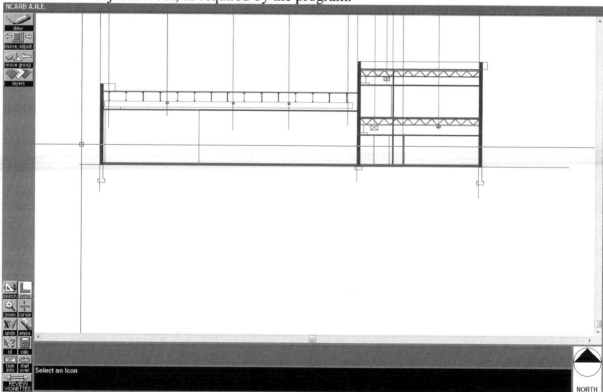

Figure 2.9   The completed solution with the sketch elements shown

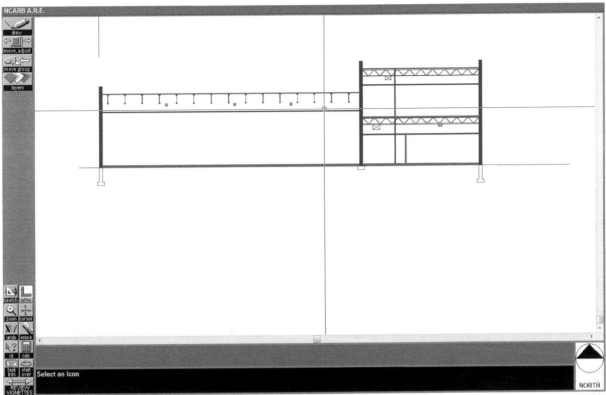

Figure 2.10   The completed solution with the sketch elements hidden

Chapter Three

ARE Mock Exam for
Construction Documents & Services (CDS) Division

A. Mock Exam: CDS Multiple-Choice (MC) Section

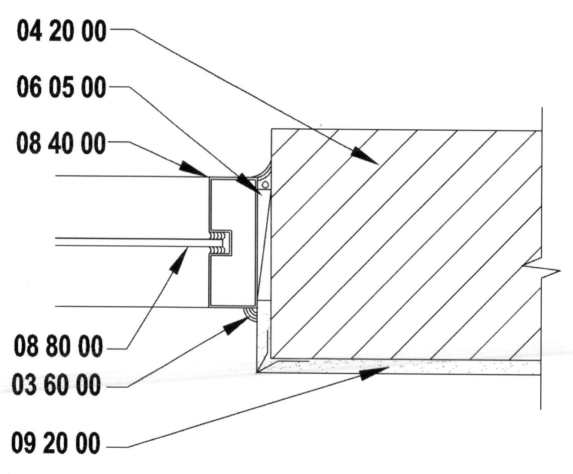

Figure 3.1   Figure for ARE CDS division mock exam MC Question #1

1. The numbered keynotes above refer to the
   a. product codes
   b. building code sections
   c. specification sections
   d. finish schedule codes

2. The occupant load factors are calculated based on
    a. the number of fixed seating
    b. the net floor area
    c. the gross floor area
    d. the building official's final decision
    e. the number of fixed seating or the gross floor area, but subject to the official's final decision
    f. the number of fixed seating, the net floor area, or the gross floor area, but subject to the official's final decision

3. For a residential remodel project, the framing and interior drywalls of a bathroom are not installed flat enough, and the mirror cannot be installed properly. The framing and interior drywalls have to be demolished and re-installed. The architect does not notice this defect in her routine job site visit. Who should pay for the extra costs?
    a. The general contractor should submit a change order to the owner to cover the extra costs.
    b. The general contractor or his subcontractors should pay for the extra costs.
    c. The architect since she does not notice this defect in her routine job site visit.
    d. The architect's professional liabilities insurance company since the architect does not notice this defect in her routine job site visit.

4. A contractor is building a shop building. The civil and plumbing site and floor plans only show the water line up to 5' (1524) outside of the building, and does not show the connecting water line from water main to within 5' of the building. There is a general note on the cover sheet of the plans instructing the contractors to construct a complete water supply system, including any connection to the water main. The contractor did not give the complete set of the plans to the plumbing subcontractor during the bidding, and the plumbing subcontractor has not included the cost for the connecting water line from water main to within 5' (1524) of the building in his bid based on the civil and plumbing site and floor plans. Who should pay for the extra costs?
    a. The general contractor since a general note on the cover sheet covers this item.
    b. The owner since the plumbing subcontractor has not included the cost for this item in the original bid.
    c. The civil engineer or the plumbing engineer since they have not properly shown this item on the civil and plumbing site and floor plans.
    d. The architect or her insurance company since she is responsible for coordinating the civil engineer or the plumbing engineer's work, and should have caught this conflict in the plans.

5. An architect is specifying the stainless steel urinals for the restrooms in a public park. She has a hard time finding the stainless steel urinals that meet both the ADA and the local building codes. What should she do?
    a. Try to talk with the building officials to gain an exception to the ADA and the local building codes based on economic hardship.
    b. Continue to look for the stainless steel urinals that meet the requirements.
    c. Use a stainless steel handicap toilet instead of the stainless steel urinals to meet the requirements and provide equal access.
    d. Install the stainless steel urinals installed in other public restrooms in the same geographic area.

6. An architect has designed the interior space for an office per codes. After the project is completed, the client thinks the counters are too low, and has his employees raise the height of the counters. The architect notices the revised counter heights do NOT comply with the building codes. She advises the client the revision does not comply with codes and should be corrected. The client refuses to accept the architect's advice and says his decision is final since he is paying the bills. What should she do?
    a. Do nothing since this is the client's property.
    b. Document the incident in a written report and update her project files.
    c. Document the incident in a written report and fax the report to the client.
    d. Report the incident to the building officials.

7. The right to use property of another without possessing it is called _____.

8. The restaurant owner informs the architect she wants to delete two toilet stalls to reduce the restroom area and increase the more profitable dining area. Prior to deleting the toilet stalls, the architect should check the requirements in:
    a. the current local building codes and plumbing codes
    b. the latest IPC
    c. the latest IBC
    d. the latest IPC and IBC

9. During the construction phase, which of the following are the architect's responsibilities? **Check the three that apply.**
    a. Review every submittal submitted by the contractor
    b. Issue changes requested by the owner
    c. Issues Addenda
    d. Field observations
    e. Issue Construction Change Directives
    f. Prepares the Punch List

10. A public school district is trying to build a new elementary school in a neighborhood. Several of the residents at the site are not willing to sell their properties. The school district can use _____ to purchase the properties at fair market value without the residents' consent.

11. A mixed-use building is \$2 million over the budget, what choices does the owner have? **Check the four that apply.**
    a. Reduce the project scope.
    b. Increase the budget by \$2 million.
    c. Ask the architect to revise the design without paying extra fees.
    d. Negotiate with the contractor to reduce the construction cost.
    e. Bill the architect's insurance company \$2 million.
    f. Back charge the architect \$2 million.

12. Which of the following statement is not true regarding substitution? **Check the three that apply.**
    a. A substitution is acceptable if it is cheaper than the specified product.
    b. A substitution can be acceptable if a specified product is not available.
    c. A substitution can be verbally approved by the owner and the architect.
    d. A substitution is acceptable if it is from a local manufacturer.
    e. A substitution cannot be submitted as part of the shop drawings without any special notation.
    f. A substitution must have equal or better quality and performance as the specified product.

13. Which of the following should be part of the written architectural service contract? **Check the two that apply.**
    a. Terms regarding reimbursable expenses
    b. Terms and conditions regarding additional services
    c. Professional liability insurance
    d. Breakdown of consultants' fee
    e. Type of the construction
    f. Schematic design fee

14. Which of the following cannot divert construction waste from reduce landfill? **Check the two that apply.**
    a. Using one dumpster at the site
    b. Using three dumpsters at the site: trash, plants and recycle
    c. Requiring all packaging materials at the site to be recycled
    d. Setting up a construction waste management plan
    e. Implementing the policy of "reduce, reuse and recycle" in the architect's office
    f. Seeking LEED certification for the building

15. Which of the following are contract documents according to A201–2007, General Conditions of the Contract for Construction? **Check the three that apply.**
    a. Invitation to bid
    b. Instruction to Bidders
    c. Supplementary Condition
    d. Contractor's proposal
    e. Drawings and Specifications
    f. Construction Change Directive

16. In contract law, the doctrine that a contract can only impose obligations or confer rights to parties to the contract is called _____.

17. Which of the followings can help assure quality control of the construction documents? **Check the three that apply.**
    a. Allowing adequate time for preparing the construction documents before bidding
    b. Having the owner review the plans
    c. Having the contractor review the plans
    d. Having a review meeting with the building officials
    e. Using checklists
    f. Using critical path method

18. Which of the following are more likely to be an architect's instrument of service? **Check the four that apply.**
    a. Civil plans
    b. Geotechnical studies
    c. Project research and studies
    d. Drawings and specifications by architect
    e. Sketches
    f. Electrical single line diagram

19. After a contract is awarded, the contractor notices that he has the site plans, but does not have the legal description of the site and the owner's interest in the site. The contractor requests the legal description from the owner. Which of the following is correct?
    a. The owner can refuse to provide the information since it is private.
    b. The contractor shall go to the county recorder's office and obtain the information since the owner is not obliged to provide it.
    c. The owner shall provide the information to the contractor.
    d. The contractor shall request the information from the civil engineer, not the owner.

20. After a project is about 50% completed, the contractor hires a new superintendent. The superintendent cannot find evidence of owner's financial arrangement for payment for the project in the contractor's project files, and she requested the owner to provide a copy of the evidence. Which of the following is correct?
    a. The owner can refuse to provide the information.
    b. The owner has to provide the information.
    c. The superintendent should ask the owner's accountant for the information.
    d. The superintendent should ask the owner's lender for the information.

21. Matters regarding the rights and responsibilities of owner, contractor and the architect to the contract regarding a specific project are best handled in:
    a. Specifications
    b. General condition
    c. Supplemental condition
    d. General notes on the cover sheet of the construction drawings

22. A project has a liquated damage cause. Because of the delay of the shipment of an electrical switchgear, the final completion is 1 month behind schedule. Who should pay the owner for monetary damages?
    a. The electrical subcontractor
    b. The contractor
    c. The supplier of the electrical switchgear
    d. The contractor should pay for most of the damages, and the electrical subcontractor should pay a portion of it.

23. An architect incorporates some specific construction methods into the construction documents per the owner's prototype plans. The contractor reviews the plans and notice the specific construction methods in the plans are not safe. He immediately calls both the owner and the architect to notify them about his concerns. The owner instructs the contractor to proceed as the original construction documents. A worker is injured while performing the work as per the construction methods in the original construction documents. Who should be responsible for this construction accident?
    a. The owner
    b. The contractor
    c. The owner and the contractor
    d. The owner and the architect

24. Who should pay for the temporary power during the construction according to A201–2007, General Conditions of the Contract for Construction?
    a. The power company
    b. The electrical subcontractor
    c. The contractor
    d. The contractor should pay for it and then get reimbursed by the owner

25. During the demolishing process of a remodel project, the contractor discovers mold after removing the gypsum board. What is the proper action for the contractor?
    a. Stop the work at the affected area, and notify the building official and the owner in writing immediately.
    b. Stop the work at the affected area, and notify the building official and the architect in writing immediately.
    c. Stop the work at the affected area, and notify the owner and the architect in writing immediately.
    d. Stop the work at the affected area, notify the owner in writing, and locate a mold specialist to solve the problem.

26. During plan check, the building official instructs the architect to add an equipment screen wall for the rooftop units (RTU) for HVAC equipment only if they are visible from two adjacent major streets by the building inspector after the RTU is installed. What is the best way for the architect to handle this in the bidding process?
   a. Ask the contractors to include the equipment screen wall as an allowance in the bid
   b. Ask the contractors to include the equipment screen wall as an alternate bid
   c. Ask the contractors to include the equipment screen wall as a change order
   d. Ask the contractors to include the equipment screen wall as an order for a minor change in the Work

27. Which of the contractor compensation methods is best for the owner if the project scope or timing is unknown, and high quality is paramount?
   a. Stipulated-sum contracts
   b. Time and materials contracts
   c. Unit-price contracts
   d. Cost-plus-fee contracts

28. During construction, the inspector requested some field tests not included in the contract documents. Who should pay for these tests?
   a. The architect since she should include all the tests required by the codes and building officials.
   b. The architect's error and omission insurance since the architect should include all the tests required by the codes and building officials.
   c. The contractor
   d. The owner

29. During construction, the owner and the contractor have a different opinion of the paint quality, what is the first step to resolve this dispute?
   a. Seeking the IDM's interpretation
   b. Mediation
   c. Litigation
   d. Arbitration

30. For a project with $10,000 allowance for a transformer, if the owner chooses to install the transformer and the actual cost of the transformer is $9,500, the contractor shall bill the owner:
   a. $9,500
   b. $10,000
   c. $9,500 plus the contractor's overhead and profit
   d. $9,500 plus the installation cost and the contractor's overhead and profit
   e. $10,000 plus the installation cost and the contractor's overhead and profit
   f. $10,000 plus the contractor's overhead and profit but not more than 15%

31. The contract documents call for 32 sets of door hardware. The contractor's submittal shows 30 sets of door hardware and is accidentally approved by the architect. The contractor later submits a claim for the extra cost for 2 sets of door hardware citing the submittal approved by the architect. What should the architect do?
    a. Negotiate with the contractor and split the extra cost of the 2 sets of door hardware.
    b. Reject the contractor's claim for the extra cost for 2 sets of door hardware.
    c. Submit a claim to the architect's error and omission insurance carrier for the extra cost for 2 sets of door hardware.
    d. Issue a change order to increase the contract sum to cover the 2 sets of door hardware.

32. At a field visit, the architect notices a subcontractor standing on a foam cornice next to the parapet. She is concerned about the safety procedures at the site. What should she do?
    a. Instruct the subcontractor not to stand on the foam cornice since it may not be strong enough to support a human weight.
    b. Do nothing, since jobsite safety procedures are the contractor's sole responsibility.
    c. Notify the owner about her concerns.
    d. Notify the superintendent about her concerns.

33. At a field visit, the architect notices the window flashings have not been installed per the contract documents. What should she do?
    a. Reject the window flashings work.
    b. Stop the work.
    c. Inform the owner and ask the owner to stop the work.
    d. Report the incident to building official.

34. The contractor unbound the plans and gave specific sections to subcontractors for bidding. The contractor later found out there are gaps in the over coverage of the bid. Which of the following statements is true?
    a. The subcontractors should split and absorb the cost of the work.
    b. The contractor should absorb the cost of the work.
    c. The owner should pay for the work because she has not paid for it in the original bid.
    d. The architect should revise supplementary condition to cover the work.

35. The contractor decided not to purchase the roof tiles at the beginning of a project because they would not be needed until months later. When he purchased the roof tiles, the price of roof tiles has risen 50% because of the damages caused by Hurricane Katrina. Which of the following is true?
    a. The contractor is not entitled to a change order.
    b. The architect should issue a change order to cover the extra cost because the damages caused by Hurricane Katrina could not be anticipated.
    c. The owner's insurance company should pay for the extra cost.
    d. The owner should pay for the extra cost.

36. Which of the following is true about the list of items to be corrected by the contractor?
    a. It should be prepared by the contractor.
    b. It should be prepared by the architect.
    c. It should be prepared by the architect, issued to the contractor, and cc the owner.
    d. It should be prepared by the architect after a joint punch walk with the contractor and the owner.

37. Which of the following construction sequence is most likely to be productive?
    a. Footing, concrete slab, framing, drywall, carpet, roofing, and painting
    b. Footing, framing, concrete slab, drywall, carpet, roofing, and painting
    c. Footing, concrete slab, framing, roofing, drywall, painting, and carpet
    d. Footing, concrete slab, framing, carpet, painting, roofing, and drywall

38. The electrical subcontractor turned off the main electrical switch for the entire floor to connect the light fixtures to the panel. The framing subcontractor did not know this. He wanted to work on the interior framing of the floor, and turned on the main electrical switch. The electrical subcontractor is killed instantly. Who is responsible to the building owner for the electrical subcontractor's death?
    a. The framing contractor
    b. The contractor
    c. The framing contractor and the contractor jointly
    d. The insurance company of the electrical contractor

39. Who should prepare the change orders?
    a. The architect
    b. The contractor
    c. The owner
    d. The superintendent

40. Which of the following inspections does an architect normally conduct?
    a. Inspection before pouring concrete for the footing
    b. Inspection before roofing
    c. Inspection before the installation of gypsum boards
    d. Inspection to determine the date of substantial completion

41. Which of the following regarding change orders and construction change directives is true? **Check the three that apply.**
    a. Change orders are for major changes, and construction change directives are for minor changes.
    b. Both change orders and construction change directives have to be signed by the owner, contractor, and the architect.
    c. When the contractor agrees and signs a construction change directive, it should be effective immediately and should be recorded as a change order.
    d. Change orders have to be signed by the owner, contractor, and the architect.
    e. Construction change directives have to be signed by the owner, contractor, and the architect.
    f. Construction change directives have to be signed by the owner and the architect.

42. When modifications to the scope of the contract occur, which of the following is true?
    a. The modifications have to be submitted to the surety for approval.
    b. The modifications do not need to be submitted to the surety for approval.
    c. The modifications have to be submitted to the owner's insurance company for approval.
    d. The modifications do not need to be submitted to the owner's insurance company for approval.

43. A contract executed on January 3, 2009 required the contractor to start the Work on January 15, 2009. The contract time is 180 days. The contractor started the Work on February 1. Substantial completion occurred on July 8, 2009. Final completion occurred on August 6, 2009. Had the contractor finished the project within the time required by the contract?
    a. No, the contractor had not finished the project within the contract time because the final completion occurred over 180 days after the commencement of the work (January 15, 2009).
    b. No, the contractor had not finished the project within the contract time because the final completion occurred over 180 days after the commencement of the work (February 1, 2009).
    c. Yes, the contractor had finished the project within the contract time because the substantial completion occurred within 180 days of the commencement of the work (January 15, 2009).
    d. Yes, the contractor had finished the project within the contract time because the substantial completion occurred within 180 days of the commencement of the work (February 1, 2009).

44. If a contractor fails to pay one of his subcontractors, the owner can issue a _____ check to the contractor and to the subcontractor who has not been paid by the contractor.

45. If asbestos or PCB is discovered at the concealed space of a project, which of the following is true? **Check the three that apply.**
    a. The contractor should stop the Work immediately.
    b. The contractor should stop the work at the affected area immediately.
    c. The contractor is entitled to a change order.
    d. The contractor is not entitled to a change order.
    e. If the contractor and the owner cannot reach an agreement regarding the asbestos issue, the contractor should submit a claim to the architect for initial determination.
    f. If the contractor and the owner cannot reach an agreement regarding the asbestos issue, the contractor should proceed to mediation.

46. Which of the following regarding boiler and machinery insurance is correct? **Check the two that apply.**
    a. It is the owner's responsibility.
    b. It is the contractor's responsibility.
    c. If there is no boiler and machinery insurance, the owner may need to pay for any related damages.
    d. If there is no boiler and machinery insurance, the contractor may need to pay for any related damages.

47. Which of the following is true regarding a waiver of subrogation in property insurance? **Check the two that apply.**
    a. It should be requested before any loss occurs.
    b. It can be requested after a loss occurs.
    c. The contractor has to disclose the subrogation provision to the insurer before purchasing the property insurance.
    d. The owner has to disclose the subrogation provision to the insurer before purchasing the property insurance.

48. When the slope of a sidewalk is more than _____% in the direction of travel, the sidewalk becomes a ramp.

49. When you are working on the construction drawings for a wood stud building, you should dimension to the _____ of the stud.

50. The five classical orders are _____.

51. Which of the following is a floor special purpose outlet?

    a.   ─⊖

    b.   ⬡

    c.   Ⓕ

    d.   △

Figure 3.2   Figure for ARE CDS division mock exam MC Question #51

1. 52. A developer is building a shopping center. All buildings in this shopping center are one-story. The occupancy group for these building is Group _____. The most cost-effective way to design and build these building is to qualify all of them as one "building," Type _____ construction, fully sprinklered, with _____ feet wide public ways or yards around all sides of the shopping center so that the area of all these buildings shall not be limited.

53. When the occupant load for a room is 780, the minimum number of exit(s) required for this room is _____.

54. Which of the following are project delivery methods? **Check the four that apply.**
    a. BIM
    b. Design-Bid-Build
    c. Construction Administration
    d. Construction Management
    e. Design-Build
    f. Integrated Project Delivery

55. Which of the following project delivery methods works best for a complicated airport project that needs to continue operating during construction?
    a. BIM
    b. Design-Bid-Build
    c. Construction Administration
    d. Construction Management
    e. Design-Build

56. If quality is the most important variable for a project, which of the following is the preferred project delivery method?
    a. Cost plus fixed fee
    b. Bridging
    c. Negotiated select team
    d. Design-Build

57. In a concrete slump test, if a shear type of slump is achieved, which of the following is true?
    a. A fresh sample should be taken and the test repeated.
    b. The test is acceptable.
    c. The test is acceptable if the shear type of slump is between 2" (51) and 5" (127).
    d. The concrete is too dry.

58. If the owner hires a design architect and a production architect, which of the following is she likely to use?
    a. Design-Bid-Build
    b. Cost plus fixed fee
    c. Bridging
    d. Negotiated select team
    e. Design-Build
    f. BIM

59. Which of the following is likely to allow the owner to know the project cost at earliest stage of the project?
    a. Design-Bid-Build
    b. Cost plus fixed fee
    c. Negotiated select team
    d. Standard design-build
    e. BIM
    f. CM-agent

60. The actual dimension of a 2x4 wood stud is _____.

61. After moving into portion of a building, the owner notices a small crack on the bathroom mirror. This crack was not on the punch (deficiency) list reviewed by the architect. What is a proper determination for the architect?
    a. The owner should be responsible for the crack since he has moved in, and the cause of the crack cannot be established.
    b. The contractor should replace the mirror even though the crack was not on the punch (deficiency) list reviewed by the architect.
    c. The architect should pay for the cost to replace the mirror because she missed the crack on the original the punch (deficiency) list walk-through.
    d. The architect's insurance company should pay for the cost to replace the mirror because the architect missed the crack on the original the punch (deficiency) list walk-through.

62. Nine months after final payment, the owner discovers the operation and maintenance manual for a 5-ton air conditioning unit was never received. What should the architect do?
    a. Request the contractor to provide the operation and maintenance manual.
    b. Withhold 5% of the retainage (holdback) until the contractor provides the operation and maintenance manual.
    c. Contact the air conditioning unit manufacturer directly to obtain the operation and maintenance manual.
    d. Have the owner contact the air conditioning unit manufacturer directly to obtain the operation and maintenance manual.

63. Which of the following is proper language for an agreement between owner and architect? **Check the two that apply.**
    a. The architect shall perform its service with professional skill and best care available in the same locality.
    b. The architect shall perform its service with professional skill and ordinary care available in the same locality.
    c. The architect shall perform its service expeditiously.
    d. The architect shall perform its service per the term of "time is of essence."

64. A statement of the company's financial activities while excluding "unusual and nonrecurring transactions" when stating how much money the company actually made is called:
    a. financial planning
    b. *pro forma* accounting
    c. revenue accounting
    d. profit/loss accounting

65. MASTERSPEC is a product of:
    a. AIA
    b. CSI
    c. ANSI
    d. ASTM

66. The U.S. National CAD Standard (NCS) promotes "Uniform Drawing Standard" (UDS). UDS uses a coordinate-based location system (grid) to organize information on each drawing sheet. Which of the following is true? **Check the two that apply.**
    a. The drawing grids are numbered from left to right in the order of 1, 2, 3...
    b. The drawing grids are numbered from right to left in the order of 1, 2, 3...
    c. The drawing grids are labeled from top to bottom in the order of A, B, C...
    d. The drawing grids are labeled from bottom to top in the order of A, B, C...

67. Per *ADAAG Manual: A Guide to the American with Disabilities Accessibility Guidelines* by Access Board, the minimum clear passage width for a single wheelchair at a doorway is _____.

68. Which of the following should occur before painting the interior? **Check the three that apply.**
    a. All interior gypsum boards are installed.
    b. Exterior windows are sealed.
    c. Exterior trees and lawn are planted.
    d. Roofing is completed.
    e. Irrigation lines are installed.
    f. Sidewalks are completed.

69. Which of the following will affect the estimated construction cost of a project? **Check the three that apply.**
    a. The experience of the contractor
    b. The experience of the architect
    c. The number of submittals
    d. The percentage of retainage (holdback)
    e. The number of RFIs
    f. The bidding climate

70. A developer wants to use local contractors to build a shopping center to help stimulate the local economic. The budget is $50 million. None of the local contractors has the bonding capacity of $50 million. Which of the following is a possible solution?
    a. Design-bid-build
    b. Fast-track
    c. Using change orders
    d. Using multiple prime contractors

71. An architect has to prepare several individual bid document packages for which of the following?
    a. Design-bid-build
    b. Design-build
    c. Bridging
    d. Fast-track

72. If an architect signs a standard B101–2007 (Former B141–1997), Standard Form of Agreement between Owner and Architect (RAIC Document 6), which of the following insurance shall she maintain during the duration of the agreement? **Check the four that apply.**
    a. General Liability
    b. Professional Liability
    c. Automobile Liability
    d. Property Liability
    e. Workers' Compensation
    f. Fire Liability

73. Per B101–2007 (Former B141–1997), Standard Form of Agreement between Owner and Architect (RAIC Document 6), which of the following is part of the architect's basic service? **Check the four that apply.**
    a. Electrical engineering
    b. Architectural design
    c. Coordination with the architect's consultants
    d. Project meetings
    e. Construction schedule
    f. Construction sequence

74. Which of the following is correct?
    a. An architect shall assist the owner and issue modifications to contractors during bidding process.
    b. An architect just needs to prepare one estimate of the construction cost to the owner.
    c. If an architect just needs to prepare estimate of the construction cost to the owner, she should treat this as an additional service.
    d. An architect shall obtain the owner's approval before proceeding to the next phase of design.

75. What is the difference between life cycle costing (LCC) and life cycle assessment (LCA)?
    a. LCA focuses on economic analysis while LCC focuses on environmental analysis.
    b. LCC focuses on economic analysis while LCA focuses on environmental analysis.
    c. LCA focuses on environmental and economic analysis.
    d. LCC focuses on environmental and economic analysis.

76. A structural engineer specifies pre-fabricated roof trusses for a project. Which of the following is correct? **Check the two that apply.**
    a. The structural engineer should show the roof trusses on the framing plans.
    b. The truss engineer should show the roof trusses on the framing plans.
    c. The structural engineer should prepare, stamp, and sign the trusses plans.
    d. The truss engineer should prepare, stamp, and sign the trusses plans.

77. During construction, a contractor submitted the roof tile substitution and followed up with a phone call the next day to the architect, requesting the architect to approve the roof tile within 2 days. The contractor also informed the architect that if the roof tile substitution was not approved within 2 days, the project would be delayed. What is a proper action for the architect?
    a. Approve the roof tile substitution within 2 days to avoid potential delay of the project.
    b. Reject the roof tile substitution.
    c. Discuss with the owner and let the owner approve or reject the substitution within a reasonable timeframe.
    d. Approve or reject the substitution within a reasonable timeframe.

78. Per B101–2007 (Former B141–1997), Standard Form of Agreement between Owner and Architect (RAIC Document 6), which of the following are additional architectural services? **Check the four that apply.**
    a. A meeting with the owner to review facility operations and performance 9 months after final completion
    b. Preparing Change Orders and Construction Change Directives and supporting documents
    c. Field Inspections to determine date of substantial completion and date of final completion
    d. Programming
    e. Building information modeling
    f. Value Analysis

79. A major fast food chain owner offers an architect an agreement between owner and architect generated by the owner's corporate attorney. The agreement designates the architect's work as "work for hire." What is the implication if the architect signs this agreement?
    a. It simply means the owner has hired the architect as a consultant for the project.
    b. It means the owner has hired the architect and obtained the license to use the architect's plans generated for the project.
    c. It means the owner has hired the architect and obtained the non-exclusive license to use the architect's plans generated for the project.
    d. It means the owner has hired the architect and obtained all the rights for the architect's plans generated for the project.

80. If a dispute arises between the owner and the architect, which of the following is a proper procedure to resolve the dispute?
    a. The owner and the architect shall seek arbitration and then mediation and then litigation.
    b. The owner and the architect shall seek mediation and then arbitration and then litigation.
    c. The owner and the architect shall seek mediation and then arbitration.
    d. The owner and the architect shall seek mediation and then arbitration or litigation.

81. Which of the following regarding mediation, arbitration, and litigation is not true?
    a. A mediator is usually a building owner, architect, contractor, or attorney practicing in the construction industry and has judge-like power.
    b. An arbiter is usually a building owner, architect, contractor, or attorney practicing in the construction industry and has judge-like power.
    c. Agreements reached in mediation and arbitration are enforceable in any court having jurisdiction thereof.
    d. Both arbitration and litigation involve a third party with judge-like power.

82. An architect can include by _____ the mechanical engineer in an arbitration with the owner regarding HVAC issues if the mechanical engineer agrees to the request in writing.

83. An owner is working with a contractor on an owner-contractor agreement, A101–2007, Standard Form of Agreement Between Owner and Contractor, where the basis of payment is a Stipulated Sum (CCDC Document 2). Which of the following is incorrect regarding revising the standard AIA form? **Check the four that apply.**
    a. Use correction fluid to wipe out the languages not used.
    b. Use Xs to completely block out the languages not used.
    c. Use single line to strike out the languages not used.
    d. Retype the AIA document in MS Word format so that it looks clean and reads better.
    e. Both parties shall initial corrections.
    f. Both parties shall initial and seal corrections.

84. A contractor received payment from the owner on March 31, 2011, up to what date did the payment cover work completed?
    a. Around March 1, 2011
    b. Around March 16, 2011
    c. Around March 24, 2011
    d. Around February 1, 2011

85. Which of the following is normally not part of the contract documents? **Check the three that apply.**
    a. General Condition
    b. Shops drawings
    c. Submittals
    d. Addenda
    e. Modifications
    f. Geotechnical reports

86. Which of the following is true regarding the final payment?
    a. The final payment is due to the contractor at substantial completion.
    b. The owner should pay the final payment 30 days after the receipt of the architect's final Certificate of Payment.
    c. If there is a surety for the project, the final payment needs the approval of the surety.
    d. The surety should issue the final Certificate of Payment.
    e. Conditional lien waivers are due before the final payment.
    f. Unconditional lien waivers for the remaining balance are due before the final payment.

87. A building official can enforce all of the following except
    a. the International Building Code
    b. the American with Disabilities Act (ADA)
    c. the Life Safety Code (NFPA 101)
    d. the National Electrical Code (NFPA 70)

88. In which CSI MasterFormat 2004 division can an architect find the specification for stone?
    a. Division 03
    b. Division 04
    c. Division 05
    d. Division 06

89. Per International Code Council, Inc. (ICC), *International Building Code* (IBC) 2006, the minimum number of plumbing fixture can be determined by: **(Check the four that apply.)**
    a. number of occupants
    b. number of rooms
    c. number of units
    d. occupancy
    e. tenant's prototype plans
    f. water pressure

90. An architect is doing initial project research. Select the three most important issues that she needs to consider from the following:
    a. zoning
    b. occupancy
    c. exit width
    d. if the building will be sprinklered
    e. plumbing fixtures
    f. HVAC equipment

91. A contractor is doing demolition work at a project. She suspects the HVAC ducts inside a concealed attic space contain asbestos. She stops the work and informs the owner. The owner hires a testing lab. The lab confirms the presence of asbestos. The project is stopped for two weeks while the owner hires a specialist to do the asbestos abatement. When the area is clear, the contractor submits a change order request for $6,000 and a time extension of two weeks. The owner rejects the change order. What is the next step?
    a. The dispute should be referred to the architect.
    b. The dispute should be referred to IDM.
    c. The dispute should proceed directly to mediation.
    d. The architect should approve the time extension, and refer the dispute of the cost to mediation.

92. An architect is working with a developer on a mixed-use complex. Which of the following is not an additional service?
    a. As-constructed record drawings
    b. Commissioning
    c. Presentations in a public hearing to defend owner's choice of water fountains and sculpture
    d. Coordination of mechanical and electrical ceiling plans

93. A form of strict, secondary liability that arises under the common law doctrine of agency, the responsibility of the superior for the acts of their subordinate is known as
    a. Employee liability
    b. Vicarious liability
    c. General liability
    d. Limited liability

94. Which of the following can affect a project's construction cost?
    I. Project location
    II. Schedule
    III. Type of construction
    IV. LEED certification

    a. I and II
    b. II and III
    c. I, II, and III
    d. All of the above

95. Which of the following items is likely to be shop fabricated?
    a. Plumbing fixtures
    b. HVAC equipment
    c. Storefront
    d. Slab

96. Which of the following statements regarding construction schedule is incorrect?
    a. The contractor is responsible for the construction schedule.
    b. The architect sets up the criteria for the construction schedule.
    c. CPM schedule shows the duration, sequences, and relationship between activities.
    d. The shortest path is the critical path in CPM.

97. An architect contracts with an electrical engineer to provide electrical engineering service for an office building. Who is responsible to the owner for the electrical system's design?
    a. The architect
    b. The electrical engineer
    c. The architect and the electrical engineer jointly
    d. The contractor

98. When selecting switchgear, an architect and his electrical engineer should consider which of the following factors? **Check the four most important factors that apply.**
    a. Dimension
    b. Voltage
    c. Noise characteristics of the equipment
    d. Clearance
    e. Available space
    f. Color

99. Which of the following are acceptable ways to revise drawings and specifications during construction? **Check the three that apply.**
    a. Addenda
    b. Notices
    c. Change orders
    d. Written orders for a minor change of the Work
    e. Construction Change directives
    f. Punch lists

100. Per C401–2007 (Former C141–1997), Standard Form of Agreement Between Architect and Consultant, which of the following statements is true?
    a. The consultant shall maintain professional liability insurance for the duration of the agreement.
    b. The consultant shall maintain workers' compensation and professional liability insurance for the duration of the agreement.
    c. The consultant shall maintain automobile liability, workers' compensation, and professional liability insurance for the duration of the agreement.

    d. The consultant shall maintain property liability, automobile liability, workers' compensation, and professional liability insurance for the duration of the agreement.

    e. The consultant shall maintain general liability, automobile liability, workers' compensation, and professional liability insurance for the duration of the agreement.

    f. There are no requirements regarding consultant's insurance

B. Mock Exam: CDS Graphic vignette section

Directions

Draw a building section according to plans and the program requirements. The section cut line indicates the location of section. Show grade line, exterior walls, bearing walls, foundations, footings, slabs on grade, finished ceilings, interior walls, ducts, joists, decks, parapets, and roofs. The section must show joists in elevation immediately adjacent to the cut line. The locations of section elements must match floor plans. The section must accurately show the minimum room heights, parapet heights, and clearance spaces needed for light fixtures and ducts. Use the minimum dimensions and clearance spaces to save building costs.

Program

The structure system for this building is open web steel joists on masonry bearing walls with continuous concrete spread footing.

1. The slab on grade is 5" (127) thick.
2. All roofs and ceilings are flat.
3. All open web steel joists have 4" (102) thick concrete deck on top.
4. The ceiling height for the multi-purpose room is 18'-0" (5486). The ceiling heights for the remaining rooms on the 1st floor are 8'-6" (2590). The ceiling heights for all the rooms on the 2nd floor are 9'-0" (2743).
5. Corridor walls are 1-hour fire-rated. Transfer grilles and fire/smoke dampers are provided as needed.
6. Exterior walls and interior bearing walls are 2-hour fire-rated.
7. The site is flat.
8. The frost depth is 4'-6" (1371).
9. The top of parapet must be 3'-0" (914) above the highest adjacent roof.
10. The clearance space for light fixtures is 8" (203). All ducts must be placed below joists, and all light fixtures must be placed below the ducts and above the ceilings.
11. All ceilings are non-rated, and the ceiling spaces are used as return air plenums.

Tips

1. Use the NCARB Practice Program to draw the section. With the help of the graphic scale on the plans, you can manually scale the horizontal dimensions from the floor plans to locate the section elements.
2. Turn **layers** on and off to become familiar with the building elements on each floor.
3. Place the elements at the approximate location, and then adjust them
4. Use **Sketch > Rectangle** to establish room heights, clearance spaces and top of parapet height.
5. Use full screen cursor to align the elements.

6. Use **Zoom** to adjust the elements to accurate locations.
7. Use **Move Group** to move the entire joist run, and **Move, Adjust** to adjust the locations of the joists within the joist run.
8. Pay attention the **Depth** and **Spacing** of the joists and the **Width** and **Depth** of the ducts when using the **Draw** command. These dimensions should match the floor plans.
9. If you cannot select one of the overlapping elements, keep clicking without moving the mouse until the desired element is highlighted.

Warning

You must meet both of the following requirements, OR you will automatically fail:
1. Draw the **grade line**.
2. The left side of your section must align with the line noted as **"EXTERIOR FACE OF SECTION"** on the plans.

Time

You must complete the vignette within 1 hour.

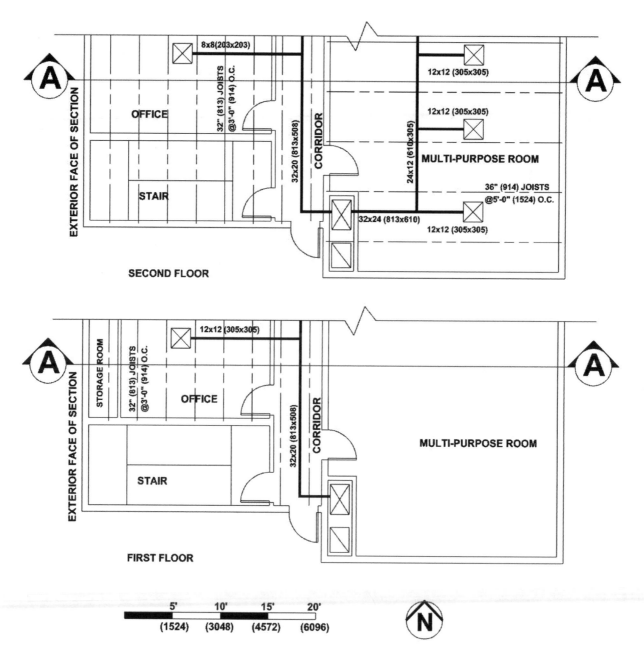

Figure 3.3 ARE CDS division mock exam: graphic vignette floor plans

Chapter Four

ARE Mock Exam Solution for
Construction Documents & Services (CDS) Division

A. Mock Exam answers and explanation: CDS Multiple-Choice (MC) Section

Note: If you answer 60% of the questions correctly, you pass the MC Section of the exam.

1. Answer: c
   These number keynotes refer to specification sections. If you are familiar with CSI MasterFormat specification sections, you should be able to answer this question correctly.

   My other book, *Building Construction*, has detailed discussions on CSI MasterFormat specification sections.

   Product codes, building code sections, and finish schedule codes are distracters.

2. Answer: f
   See IBC, Section 1004 and Table 1004.1.1.
   The occupant load factors are calculated based on the number of fixed seating, the net floor area, or the gross floor area, but subject to the official's final decision. For example, per Table 1004.1.1, the occupant load factor for mercantile use is based on the gross floor area, but the occupant load factor for day care use is based on the net floor area. See Section 1004.1.1 for examples of exceptions that a building official can grant.

3. Answer: b
   It is the contractor's responsibility to coordinate all his subcontractors' work and make sure the construction work is performed per the contract documents. The owner has nothing to do with this. The architect's routine job site visits are for field observation and to verify the construction work is in general compliance with the contract documents, but it is still the contractor's responsibility to coordinate all his subcontractors' work and make sure the construction work is performed per the contract documents.

4. Answer: a
   The contractor will pay for the missed item since a general note on the cover sheet covers this item. Per A201–2007, General Conditions of the Contract for Construction, the contract documents are complementary, and what is required by one shall be as binding as if required by all. The contractor should have coordinated with all his subcontractors and submitted the bid based on the entire set of contract documents, not just one

On the other hand, as an architect, you should also coordinate with your consultants to avoid conflicts within the contract documents. This shows you the importance of coordinating the plans and specifications.

5. Answer: b
    She should continue to look for the stainless steel urinals that meet the requirements.
    She can try to talk with the building officials to gain an exception to the ADA and the local building codes, but this request should NOT be based on economic hardship. Using a stainless steel handicap toilet instead of the stainless steel urinals will NOT really provide equal access. The stainless steel urinals installed in other public restrooms in the same geographic area may NOT meet the requirements.

6. Answer: d
    According to Rule 3.3 of the *Rules of Conduct* published by NCARB, she should report the violation to the building officials.

    The *Rules of Conduct* is available as a FREE PDF file at:
    http://www.ncarb.org/

7. Answer:
    The right to use property of another without possessing it is called **Easement**.

8. Answer: a
    Prior to deleting the toilet stalls, the architect should check the requirements in the current local building codes and plumbing codes. The IPC and IBC are NOT enforceable until they are adapted as part of the local codes. The governing agencies need some time to adapt the IPC or IBC, and the latest IPC or IBC may NOT be have been adapted yet.

9. Answer: b, d, and e
    The architect does NOT need review **every** submittal submitted by the contractor. She just needs to review the submittals required by the construction documents. Other submittals not required by the construction documents submitted by the contractor can be returned without ANY action by the architect.

    The architect issues Addenda before the bid is due, NOT during construction phase.

    The contractor, NOT the architect, prepares the Punch List.

    The architect conducts field observations and issues changes requested by the owner via Construction Change Directives or Change Orders during construction phase.

10. Answer:
    The public school district can use **condemnation via eminent domain (or expropriation)** to purchase the properties at fair market value without the residents' consent.

11. Answer: a, b, c, and d

Per B101–2007 (Former B141–1997), Standard Form of Agreement Between Owner and Architect (RAIC Document 6), the owner can reduce the project scope, increase the budget by $2 million, ask the architect to revise the design without paying extra fees, and negotiate with the contractor to reduce the construction cost. She cannot bill the architect's insurance company $2 million, or back charge the architect $2 million.

12. Answer: a, c, and d

Please pay attention to the word "not." Since we are looking for the statements that are NOT true, the following <u>untrue</u> statements are the <u>correct answers</u>:
- A substitution is acceptable if it is cheaper than the specified product.
- A substitution can be verbally approved by the owner and the architect.
- A substitution is acceptable if it is from a local manufacturer.

The following <u>true</u> statements are the <u>incorrect answers</u>:
- A substitution can be acceptable if a specified product is not available.
- A substitution cannot be submitted as part of the shop drawings without any special notation.
- A substitution must have equal or better quality and performance as the specified product.

13. Answer: a and b

Terms and condition regarding reimbursable expenses and additional services should be part of the written architectural service contract. The other items such as professional liability insurance, type of the construction, and schematic design fee may be part of some written architectural service contracts, but they may not even be needed for other cases.

14. Answer: a and e

Please pay attention to the word "not." Since we are looking for the measures that can<u>not</u> divert construction waste from landfill, the following are the <u>correct answers</u>:
- Using one dumpster at the site
- Implementing the policy of "reduce, reuse, and recycle" in the architect's office (this will reduce waste, but NOT construction waste)

The following measures are the <u>incorrect answers</u>:
- Using three dumpsters at the site: trash, plants, and recycle
- Requiring all packaging materials at the site to be recycled
- Setting up a construction waste management plan
- Seeking LEED certification for the building

15. Answer: c, e, and f

Per A201–2007, General Conditions of the Contract for Construction, the following are contract documents:
- Supplementary Condition
- Drawings and Specifications
- Construction Change Directive

The following are NOT contract documents unless specifically enumerated in the agreement:
- Invitation to bid
- Instruction to Bidders
- Contractor's proposal

You can find a FREE PDF file of commentary for AIA document A201–2007, General Conditions of the Contract for Construction, at the following link:
http://www.aia.org/contractdocs/aiab081438

16. Answer:
In contract law, the doctrine that a contract can only impose obligations or confer rights to parties to the contract is called **privity**.

See the FREE PDF file of commentary for AIA document A201–2007, General Conditions of the Contract for Construction.

17. Answer: a, c, and e
The following can help assure quality control of the construction documents, and they are the correct answers:
- Allowing adequate time for preparing the construction documents before bidding (This will definitely help assure quality control of the construction documents)
- Having the contractor review the plans (The contractor is an expert in construction and can really help assure quality control of the construction documents)
- Using checklists (A very efficient way of assure quality control of the construction documents)

The following are the incorrect answers:
- Having the owner review the plans (Owner is not an expert in construction and cannot really help assure quality control of the construction documents)
- Having a review meeting with the building officials (Building officials may help you in interpreting building codes, but they are not expert in construction and cannot really help assure quality control of the construction documents)
- Using critical path method (This is part of the fast-track method, and has nothing to do with assuring quality control of the construction documents)

18. Answer: c, d, e, and f
Architect's instrument of service includes every embodiment of the professional services that architect and his consultants provide, including project research and studies, sketches, drawings, and specifications. Structural, electrical, mechanical, and plumbing engineers are typically the architect's consultants, and their plans and specifications are part of the architect's instrument of service. Electrical single line diagram is part of the electrical plans and therefore part of architect's instrument of service.

Civil and geotechnical engineers are often owner's consultants, and civil plans and geotechnical studies are unlikely to be part of architect's instrument of service unless noted on the plans.

19. Answer: c

The owner shall provide the information to the contractor within 15 days of receiving the request per A201–2007, General Conditions of the Contract for Construction.

20. Answer: a

Per A201–2007, General Conditions of the Contract for Construction, after the commencement of the work, the owner does NOT have to provide evidence of owner's financial arrangement for the project unless:

- The owner fails to make the payments per the contract documents
- A project's scope change affects the contract sum
- The contractor indicates in writing a reasonable concern regarding owner's abilities to make payment when due.

21. Answer: c

Matters regarding the rights and responsibilities of owner, contractor and the architect to the contract regarding a <u>specific</u> project are best handled in <u>supplemental condition</u>.

22. Answer: b

The contractor should pay the owner for monetary damages. The contractor can back charge his electrical subcontractor and the supplier of the electrical switchgear to recover the cost, but that is a separate issue.

The contractor has to pay the owner for monetary damages per the contract even if he may not be able to recover his cost.

According to **privity**, the electrical subcontractor and the supplier of the electrical switchgear are not a party to the contract to the owner, and they should not pay the owner directly.

23. Answer: b

The contractor should be responsible for this construction accident. If he has notified the owner and the architect to about his concerns <u>in writing instead of verbally</u>, and the owner instructs the contractor to proceed as the original construction documents, the owner would be responsible for the accident. See A201–2007, General Conditions of the Contract for Construction, Section 3.3.1.

24. Answer: c

The contractor should pay for the temporary power during the construction according to A201–2007, General Conditions of the Contract for Construction, Section 3.4.1.

25. Answer: c

   The proper action for the contractor is to stop the work at the affected area and notify the owner and the architect in writing immediately, but in no event later than 21 days after the first observance of the concealed condition. See A201–2007, General Conditions of the Contract for Construction, Section 3.7.4 and 10.3.2.

26. Answer: a

   The best way to handle this in the bidding process is to ask the contractors to include the equipment screen wall as an allowance in the bid. See A201–2007, General Conditions of the Contract for Construction, Section 3.8. A change order and an order for a minor change in the Work are used during construction, not the bidding process.

27. Answer: d

   The cost-plus-fee contract is best for the owner if the project scope or timing is unknown, and high quality is paramount.

   See *The Architect's Handbook of Professional Practice* (AHPP).

28. Answer: d

   The owner should pay for the field tests not included in the contract documents.

29. Answer: a

   During construction, the first step to resolve the dispute between the owner and the contractor is seeking the Initial Decision Maker's (IDM's) interpretation. The IDM is typically the architect unless the owner and the contractor designated another party to be the IDM in the contract documents.

   See FREE PDF file of commentary for AIA document A201–2007, General Conditions of the Contract for Construction.

30. Answer: a

   The architect will issue a change order to adjust the contract sum to reflect the $500 reduction from the allowance amount.

   The contractor shall bill the owner $9,500. The installation cost, the contractor's overhead, and profit are ALREADY included in the contract sum.

   See the FREE PDF file of commentary for AIA document A201–2007, General Conditions of the Contract for Construction.

31. Answer: b

   Submittals are NOT a part of the contract documents. The architect's act of approving submittals does NOT relieve the contractor from its obligation of complying with the contract documents.

See FREE PDF file of commentary for AIA document A201–2007, General Conditions of the Contract for Construction.

32. Answer: d

    Jobsite safety procedures are the contractor's sole responsibilities, but the architect should still notify the superintendent about her concerns. She should never instruct a subcontractor what to do. Notifying the owner may be helpful, but it is not the best answer.

33. Answer: a

    Reject the window flashings work. Rejecting work is a primary means for the architect to request the contractor to correct defects or deficiencies in the contractor's work. It is also the best choice for this situation.

    Owner, not the architect, has the authority to stop work.

    See FREE PDF file of commentary for AIA document A201–2007, General Conditions of the Contract for Construction.

34. Answer: b

    The contractor should absorb the cost of the work. The contractor should submit bid based on the entire set of contract documents, NOT just portions. It is the contractor's responsibility to coordinate all subcontractors and submit a bid for the entire Work.

35. Answer: a

    The contractor is not entitled to a change order. The project has started, and the contract must have been signed. The contractor must fulfill his responsibilities under the contract.

36. Answer: a

    The punch list (the list of items to be corrected by the contractor) should be prepared by the contractor.

37. Answer: c

    The construction sequence of footing, concrete slab, framing, roofing, drywall, painting, and carpet is most likely to be productive.

    The key is roofing must happen before drywall, painting, and carpet so that the roof can protect these indoor work from rain or other weather conditions.

38. Answer: b

    The contractor is responsible to the building owner for the electrical subcontractor's death since the contractor is responsible to the owner for the all his work and his subcontractors' work and the safety procedures at the job site. The building owner has a contract with the contractor, NOT the subcontractors. However, the subcontractors are responsible to the contractor for their portion of the work.

39. Answer: a

The architect should prepare the change orders.

40. Answer: d

An architect normally conducts the following inspections:
- Inspection to determine the date or dates of substantial completion
- Inspection to determine the date of final completion

The other answers are just **distracters**.

41. Answer: c, d, and f

The following regarding change orders and construction change directives are true:
- When the contractor agrees and signs a construction change directive, it should be effective immediately and should be recorded as a change order.
- Change orders have to be signed by the owner, contractor, and the architect.
- Construction change directives have to be signed by the owner and the architect.

The following regarding change orders and construction change directives are untrue:
- Change orders are for major changes, and construction change directives are for minor change. (Both can be for major changes.)
- Both change orders and construction change directives have to be signed by the owner, contractor, and the architect. (Construction change directives do NOT have to be signed by the contractor.)
- Construction change directives have to be signed by the owner, contractor, and the architect.

See the FREE PDF file of commentary for AIA document A201–2007, General Conditions of the Contract for Construction.

42. Answer: a.

The modifications have to be submitted to the surety for approval. Per AIA Document A312-1984, article 8: "The Surety hereby waives notice of any change, including changes of time, to the Construction Contract or to related subcontracts, purchase orders and other obligations." However, A312-1984, article 8 does NOT prevent modifications from being submitted to the surety for approval. Per the AIA Document **A201-2007** Commentary, Section 7.3.1: "Modifications that materially alter the scope of the contract **should be** submitted for approval of the surety **to ensure** that the surety will not be released from its obligations by such changes." So, based on all the facts so far, the correct answer is a.

A surety is the contractor's insurance company since the contractor is paying for the premium, even though the owner is named as the insured. It is different from the owner's insurance company.

43. Answer: c

Yes, the contractor had finished the project within the contract time because the substantial completion occurred within 180 days of the commencement of the work (January 15, 2009).

The date of the commencement of the work is the date established in the agreement, NOT the actual date the contractor start the work.

Contract time ends on the date of the substantial completion, NOT the date of the final completion.

44. Answer:

If a contractor fails to pay one of his subcontractors, the owner can issue a **joint check** to the contractor and to the subcontractor who has not been paid by the contractor.

A **joint check** is a check made payable to two parties, and neither payee can cash or deposit the **joint check** without the endorsement or consent of the other.

45. Answer: b, c, and f

If asbestos or PCB or other hazardous materials are discovered at the concealed space of a project, the contractor:
- should stop the work at the affected area immediately
- is entitled to a change order.
- should proceed to mediation directly for disputes (Disputes related to hazardous materials should proceed to mediation directly. They are different from most other claims that are normally submitted to the architect for initial determination).

46. Answer: a and c

Boiler and machinery insurance is the owner's responsibility. If there is no boiler and machinery insurance, and the owner does not notice that the contractor does not intend to purchase the boiler and machinery insurance, then the owner effectively becomes the insurer, and she will need to pay for any related damages.

47. Answer: a and d

A waiver of subrogation should be requested before any loss occurs. The owner has to disclose the subrogation provision to the insurer before purchasing the property insurance.

48. When the slope of a sidewalk is more than **5%** in the direction of travel, the sidewalk becomes a ramp.

49. When you are working on the construction drawings for a wood stud building, you should dimension to the **face** of stud.

50. The five classical orders are **Tuscan, Doric, Ionic, Corinthian, and Composite.**

You should have learned about them in architectural history classes. You can also find this information in *Architectural Graphic Standards*.

51. Answer: d
    Answer "a" is the symbol for single receptacle outlet, answer "b" is the symbol for floor single receptacle outlet, answer "c" is the symbol for fan hanger receptacle, and answer "d" is the correct answer: the symbol for floor special purpose outlet.

    See *Architectural Graphic Standards*.

52. A developer is building a shopping center. All buildings in this shopping center are one-story. The occupancy group for these building is Group M. The most cost-effective way to design and build these building is to qualify all of them as one "building," Type V construction, fully sprinklered, with 60-feet-wide public ways or yards around all sides of the shopping center so that the area of all these buildings shall not be limited.

    See Section 507 and Table 1004.1.1 of *International Building Code* (IBC) 2006, by International Code Council, Inc. (ICC).

    See the following link for some FREE IBC code sections citations:
    http://publicecodes.citation.com/icod/ibc/2006f2/index.htm?bu=IC-P-2006-000001&bu2=IC-P-2006-000019

53. When the occupant load for a room is 780, the minimum number of exit(s) required for this room is 3.

    See Table 1019.1 of *International Building Code* (IBC) 2006, by International Code Council, Inc. (ICC).

54. Answer: b, d, e, and f
    The following are project delivery methods:
    • Design-Bid-Build
    • Construction Management
    • Design-Build
    • Integrated Project Delivery

    BIM is an enabling technology, NOT a project delivery method.

    Construction Administration is one stage in construction, NOT a project delivery method. Both BIM and Construction Administration are distracters to confuse you.

    See *The Architect's Handbook of Professional Practice* (AHPP).

55. Answer: e
    Design-Build works best for a complicated airport project that needs to continue operating during construction.

    See *The Architect's Handbook of Professional Practice* (AHPP).

56. Answer: c
   If quality is the most important variable for a project, negotiated select team is the preferred project delivery method.

   Cost plus fixed fee is the preferred project delivery method if the project scope is not defined clearly.

   Design-Build is the preferred project delivery method if the project risk is the most important variable for a project. Bridging is a sub-category of design-build.

   See *The Architect's Handbook of Professional Practice* (AHPP).

57. Answer: a
   In a concrete **slump test**, if a shear slump or collapse slump is achieved, the concrete is too wet. The test is NOT acceptable, and a fresh sample should be taken and the test repeated.

   See the following link for more information:
   http://en.wikipedia.org/wiki/Concrete_slump_test

58. Answer: c
   If the owner hires a design architect and a production architect, she is likely to use bridging or bridge design-build.

   Design-Bid-Build is a traditional project delivery method, and often involves one architect.

   Design-Build involves an entity that does both design and construction, and often involves one architect.

   If quality is the most important variable for a project, negotiated select team is the preferred project delivery method. Negotiated select team often involves one architect.

   BIM is an enabling technology, NOT a project delivery method.

   Cost plus fixed fee is the preferred project delivery method if the project scope is not defined clearly. Cost plus fixed fee often involves one architect.

   Design-build includes standard Design-build and bridging. It is a possible answer, but it is NOT as good as bridging or bridge design-build.

   See *The Architect's Handbook of Professional Practice* (AHPP).

59. Answer: d
   Standard design-build is likely to allow the owner to know the project cost at earliest stage of the project: before the Predesign (PD) stage.

Design-Bid-Build is likely to allow the owner to know the project cost after the Contract Documents (CD) are completed.

Cost plus fixed fee and negotiated select team are likely to allow the owner to know the project cost after the Schematic Design (SD) stage.

BIM is an enabling technology, NOT a project delivery method.

CM-agent is likely to allow the owner to know the project cost after the Contract Documents (CD) are completed.

See *The Architect's Handbook of Professional Practice* (AHPP).

60. Answer:
    The actual dimensions of a 2x4 wood stud are   1 ½" x 3 ½" (38x89).

    The actual dimensions of **wood** studs are **smaller** than the nominal dimensions, but the actual dimensions of **metal** studs **match** the nominal dimensions. This difference is important for coordination. For example, the actual dimension of a 4" metal stud is 4" (102). Your structural engineer can place a 4x4 (102x102) tube steel column inside a 4" (102) metal stud wall, but she can NOT place it inside a 2x4 wood stud wall since its actual dimension is 1 ½" x 3 ½" (38x89). When you review your structural engineer's foundation plans and framing plans, you need to pay attention to issues like this.

    See the following link for more information:
    http://en.wikipedia.org/wiki/Lumber

61. Answer: a
    The owner should be responsible for the crack since he has moved in, and the cause of the crack cannot be established.

62. Answer: a
    The architect should request the contractor to provide the operation and maintenance manual since this is the contractor's responsibility.

    Contacting the air conditioning unit manufacturer directly to obtain the operation and maintenance manual, or having the owner contact the air conditioning unit manufacturer directly to obtain the operation and maintenance manual are possible solutions, but they are not as good as the solution above.

    This occurs nine months after final payment. There is no money left in the retainage (holdback).

63. Answer: b and c
    The following are proper language for an agreement between owner and architect:

- The architect shall perform its service with professional skill and **ordinary** care available in the same locality.
- The architect shall perform its service expeditiously.

The following are improper language for an agreement between owner and architect:
- The architect shall perform its service with professional skill and **best** care available in the same locality. (This statement will put the architect in a disadvantage position and bring on unnecessary liabilities because it takes a lot more than average time and effort to achieve "best" result. "Best" is also a very subjective term that can be interpreted differently by different people.)
- The architect shall perform its service per the term of "time is of essence." (This statement will put the architect in a disadvantage position and bring on unnecessary liabilities and requires the architect to meet absolute time limitation. There are factors that cannot be controlled by the architect, such as the building department plan check time, etc. A written schedule may be helpful, but it must be subject to changes that are beyond the control of the owner and the architect.)

See the FREE PDF file of the commentary for AIA documents B101 at the following link for more information:
http://www.aia.org/contractdocs/aiab081438

64. Answer: b
A statement of the company's financial activities while excluding "unusual and nonrecurring transactions" when stating how much money the company actually made is called *pro forma* **accounting**. All other answers are distracters.

See *The Architect's Handbook of Professional Practice* (AHPP), Section 8.4, or the following link for more information:
http://en.wikipedia.org/wiki/Pro_forma

65. Answer: a
MASTERSPEC is a product of AIA, published and supported by ARCOM.

MasterFormat is a product of CSI.

MASTERSPEC uses assigned numbers per MasterFormat. There are related but different.

ANSI and ASTM are distracters.

See *The Architect's Handbook of Professional Practice* (AHPP), Section 12.3, or the following link for more information:
http://www.arcomnet.com/masterspec/index.php

66. Answer: a and d
The drawing grids are numbered from <u>left</u> to <u>right</u> in the order of 1, 2, 3...

AND

The drawing grids are labeled from <u>bottom</u> to <u>top</u> in the order of A, B, C...

See *The Architect's Handbook of Professional Practice* (AHPP), Section 12.3, or the following links for more information:
http://www.NationalCADStandard.org

OR
http://www.buildingsmartalliance.org/index.php/ncs/content/

67. Answer:
Per *ADAAG Manual: A Guide to the American with Disabilities Accessibility Guidelines* by Access Board, the minimum clear passage width for a single wheelchair at a doorway is <u>32"(813)</u>.

See the following link for more information:
http://www.access-board.gov/adaag/html/figures/fig1.html

You need to become familiar with all the Accessibility diagrams and critical dimensions at the following link:
http://www.access-board.gov/adaag/html/figures/index.html

68. Answer: a, b, and d
The following should occur before painting interior (The explanations are in parentheses below. Typical for the rest of the answers):
- All interior gypsum boards are installed.
- Exterior windows are sealed (This will prevent water or moisture damage if it rains).
- Roofing is completed (This will prevent water damage if it rains).

The following can occur after painting interior:
- Exterior trees and lawn are planted.
- Irrigation lines are installed.
- Sidewalks are completed.

69. Answer: a, b, and f
The following will affect the estimated construction cost of a project:
- The experience of the contractor
- The experience of the architect (An architect with good experience can select better building systems, produce high quality plans and specifications, and reduce the construction cost of a project.)
- The bidding climate (It means the anticipated weather conditions during the duration of a project.)

The following will not affect the estimated construction cost of a project:
- The number of submittals (The contractor shall construct the project per the contract documents, even if some of the items do not require submittals.)

- The percentage of retainage (holdback)
- The number of RFIs (Some contractors tend to send in many Requests for Information, or RFIs, and hope to turn some of the RFIs into change orders. If an item is already covered in the contract documents, then it will NOT be a change order item even if it is on the RFI. It is important for an architect to tie the RFI responses back to the contract documents to reduce potential change orders. The contractor shall construct the project per the contract documents.)

70. Answer: d
    Using multiple prime contractors is a possible solution. The other three choices will NOT solve the problem.

71. Answer: d
    An architect has to prepare several individual bid document packages for fast-track projects, such as a civil bid document package, foundation and superstructure bid document package, building enclosure bid document package, interiors bid document package, etc.

    The other three delivery methods (design-bid-build, design-build, including bridging) have one construction contract.

    See *The Architect's Handbook of Professional Practice* (AHPP) for more information.

72. Answer: a, b, c, and e
    If an architect signs a standard B101–2007 (Former B141–1997), Standard Form of Agreement between Owner and Architect (RAIC Document 6), she shall maintain the following insurance during the duration of the agreement:
    - General Liability
    - Professional Liability
    - Automobile Liability
    - Workers' Compensation

    See the FREE PDF file of the commentary for B101–2007 (Former B141–1997), Standard Form of Agreement between Owner and Architect (RAIC Document 6) at the following link:
    http://www.aia.org/contractdocs/aiab081438

73. Answer: a, b, c, and d
    Per B101–2007 (Former B141–1997), Standard Form of Agreement between Owner and Architect (RAIC Document 6), an architect's **basic** services include the following:
    - Electrical, Mechanical, and Structural engineering (An architect can hire engineers as sub-consultants to perform the work.)
    - Architectural design
    - Coordination with the architect's consultants
    - Project meetings
    - Complying with the design requirements of the utilities companies and governing agencies

- Feasibility of incorporating environmentally responsible design

An architect's basic services do NOT include the following:
- Construction schedule (Part of the contractor's work)
- Construction sequence (Part of the contractor's work)

See Article 3, Scope of the Architect's Basic Services in the FREE PDF file of the commentary for B101–2007 (Former B141–1997), Standard Form of Agreement between Owner and Architect (RAIC Document 6) at the following link: http://www.aia.org/contractdocs/aiab081438

74. Answer: d

An architect shall obtain the owner's approval before proceeding to the next phase of design.

The following three answers are distracters, and they are not correct:
- An architect shall assist the owner and issue modifications to contractors during bidding process. (An architect shall assist the owner and issue **Addenda**, NOT **modifications,** to contractors during bidding process, and **before** the execution of the construction contract. **Modifications** only occur **after** the execution of the construction contract, in the form of change orders, construction change directives, or an order for minor change in work. Make sure you know the difference between the Addenda and Modifications.)
- An architect just needs to prepare one estimate of the construction cost for the owner. (An architect needs to prepare estimates or updates of the estimates of the construction cost for the owner at the preliminary design, schematic design, design development, and construction documents phase as part of the **basic** architectural services)
- If an architect just needs to prepare an estimate of the construction cost for the owner, she should treat this as an additional service (See explanation for the previous item).

See the FREE PDF file of the commentary for B101–2007 (Former B141–1997), Standard Form of Agreement between Owner and Architect (RAIC Document 6) at the following link:
http://www.aia.org/contractdocs/aiab081438

75. Answer: b

LCC focuses on economic analysis while LCA focuses on environmental analysis.

See *The Architect's Handbook of Professional Practice* (AHPP) for LCC.

76. Answer: a and d

The following is correct:
- The structural engineer should show the roof trusses on the framing plans.
- The truss engineer should prepare, stamp, and sign the trusses plans. (Trusses are typically design-build by the truss engineer retained by the contractor.)

See the FREE PDF file of the commentary for B101–2007 (Former B141–1997), Standard Form of Agreement between Owner and Architect (RAIC Document 6) at the following link:
http://www.aia.org/contractdocs/aiab081438

77. Answer: c
   Discuss with the owner and have the owner approve or reject the substitution within a reasonable timeframe. The owner, NOT the architect, has the right to approve or reject the substitution. If the owner approves the substitution, the architect can issue an order for minor changes in the Work, a Change Order, or Construction Change Directive regarding this substitution.

   In most cases, the contractor requests substitutions because they are cheaper; an architect should review the request and request proper credit from the contractor if applicable.

   The roof tiles are a long-lead item, and the contractor should have ordered them well in advance. Any delay because of the substitution request should be the responsibility of the contractor.

   See the FREE PDF file of commentary for AIA document A201–2007, General Conditions of the Contract for Construction, for related information.

78. Answer: b, d, e, and f
   Per B101–2007 (Former B141–1997), Standard Form of Agreement between Owner and Architect (RAIC Document 6), the following are **additional** architectural services:
   - Preparing Change Orders and Construction Change Directives and supporting documents (Supporting documents are provided by the contractor, the architect can provide them as an additional architectural service)
   - Programming
   - Building information modeling
   - Value Analysis

   The following are **basic** architectural services:
   - A meeting with the owner to review facility operations and performance 9 months after final completion
   - Field Inspections to determine date of substantial completion and date of final completion

   See Article 3 and Article 4 of the FREE PDF file of the commentary for B101–2007 (Former B141–1997), Standard Form of Agreement between Owner and Architect (RAIC Document 6) at the following link:
   http://www.aia.org/contractdocs/aiab081438

79. Answer: d
   In the US, the term "work for hire" means the owner has hired the architect and obtained ALL the rights for the architect's plans and other documents generated for the project.

Once an architect transfers all the rights for plans and other documents generated for a project, she cannot even re-use them for another project herself. Make sure you know the implication when you review any contract that deviates from the AIA documents.

The standard B101–2007 (Former B141–1997), Standard Form of Agreement between Owner and Architect (RAIC Document 6) clearly defines that all plans and other documents are instrument of service and the architect owns all the rights and grants a non-exclusive license right (NOT copyright) for the owner to use them for this project ONLY. If the owner wants to use the plans for another project, he needs pay the architect an additional license fee and release and indemnify the architect for such use.

See Article 7 of the FREE PDF file of the commentary for B101–2007 (Former B141–1997), Standard Form of Agreement between Owner and Architect (RAIC Document 6) at the following link:
http://www.aia.org/contractdocs/aiab081438

80. Answer: d
The owner and the architect shall seek mediation, and then arbitration or litigation.

See Article 8 of the FREE PDF file of the commentary for B101–2007 (Former B141–1997), Standard Form of Agreement between Owner and Architect (RAIC Document 6) at the following link:
http://www.aia.org/contractdocs/aiab081438

81. Answer: a
A mediator is usually a building owner, architect, contractor, or attorney practicing in the construction industry and has judge-like power. (A mediator does **NOT** have judge-like power and can**not** impose settlement terms.)

Please note we are looking for the **incorrect** statement as the correct answer.

The following are **correct** statements, or the **incorrect** answers:
• An arbiter (or arbitrator) is usually a building owner, architect, contractor. or attorney practicing in the construction industry and has judge-like power.
• Agreements reached in mediation and arbitration are enforceable in any court having jurisdiction thereof.
• Both arbitration and litigation involve a third party with judge-like power.

See Article 8 of the FREE PDF file of the commentary for B101–2007 (Former B141–1997), Standard Form of Agreement between Owner and Architect (RAIC Document 6) at the following link:
http://www.aia.org/contractdocs/aiab081438

82. Answer:
An architect can include by **joinder** the mechanical engineer in an arbitration with the owner regarding HVAC issues if the mechanical engineer agrees to the request in writing.

See Article 8 of the FREE PDF file of the commentary for B101–2007 (Former B141–1997), Standard Form of Agreement between Owner and Architect (RAIC Document 6) at the following link:
http://www.aia.org/contractdocs/aiab081438

83. Answer: a, b, d, and f

Please note we are looking for the **incorrect** statement as the correct answer.

The following are incorrect statements, or the correct answers:
- Use correction fluid to wipe out the languages not used. (No, it may raise suspicion of fraudulent concealment.)
- Use Xs to completely block out the languages not used. (No, it may raise suspicion of fraudulent concealment.)
- Retype the AIA document in MS Word format so that it looks clean and reads better. (No, it can introduce typographic errors and cloud legal interpretation of a standard clause, and it is a copyright infringement.)
- Both parties shall initial and seal corrections. (No, no need to seal.)

The following are correct statements, or the incorrect answers:
- Use single line to strike out the languages not used.
- Both parties shall initial corrections.

See Instructions for A101–2007, Standard Form of Agreement Between Owner and Contractor where the basis of payment is a Stipulated Sum (CCDC Document 2), which of the following is incorrect regarding revising the standard AIA form.

84. Answer: b
The payment covered work completed up to or around March 16, 2011, because it takes the architect around 7 days to review the application, and the owner around 7 days to pay the contractor.

See Article 5.1.3, Instructions for A101–2007, Standard Form of Agreement Between Owner and Contractor where the basis of payment is a Stipulated Sum (CCDC Document 2).

85. Answer: b, c, and f

Please note we are looking for a document that is normally **not** part of the contract documents as the correct answer.

The following are normally not part of the contract documents, or the correct answers:

- Shops drawings
- Submittals
- Geotechnical reports

The following are normally part of the contract documents, or the incorrect answers:
- General Condition
- Addenda
- Modifications

See A101–2007, Standard Form of Agreement Between Owner and Contractor where the basis of payment is a Stipulated Sum (CCDC Document 2), which of the following is incorrect regarding revising the standard AIA form.

86. Answer: c and e.
The following are true regarding the final payment:
- If there is a surety for the project, the final payment needs the approval of the surety.
- Conditional lien waivers are due before the final payment.

The following are not true regarding the final payment:
- The final payment is due to the contractor at <u>substantial</u> completion. (The final payment is due to the contractor at the <u>final</u> completion.)
- The owner should pay the final payment 30 days <u>after</u> the <u>receipt</u> of the Architect's final Certificate of Payment. (The owner should pay the final payment <u>no later than</u> 30 days after the **issuance** of the Architect's final Certificate of Payment.)
- The surety should issue the final Certificate of Payment. (The architect should issue the final Certificate of Payment.)
- Unconditional lien waivers for the remaining balance are due before the final payment. (Unconditional lien waivers for the remaining balance are due <u>after</u> the final payment.)

87. Answer: b
Pay attention to the word "except": A building official canNOT enforce the American with Disabilities Act (ADA) because it is a civil rights registration, NOT a building code. ADA can be a base for civil litigation, and is enforced only by the Department of Justice (DOJ). The other choices are model codes that can be adopted by a local governing agency to become building codes.

A building official can enforce all of the following:
- the International Building Code
- the Life Safety Code (NFPA 101)
- the National Electrical Code (NFPA 70)

88. Answer: b
An architect finds the specification for stone in which Division 04 of MasterFormat 2004 division.

Per MasterFormat 2004, the divisions mentioned in this question are as follows:

- Division 03   Concrete
- Division 04   Masonry (Including 04 50 00 Stone)
- Division 05   Metals
- Division 06   Wood, Plastics, and Composites

You should become familiar with MasterFormat 2004. My other book, *Building Construction*, has detailed discussions on CSI MasterFormat specification sections.

89. Answer: a, b, c, and d
Per International Code Council, Inc. (ICC), *International Building Code* (IBC) 2006, the minimum number of plumbing fixture can be determined by:
- number of occupants
- number of rooms (See Institutional Occupancy Group)
- number of units (See Residential Occupancy Group)
- occupancy

Tenant's prototype plans can determine the minimum number of plumbing fixtures for a project, but it is NOT a part of IBC. Water pressure is just a distracter.

See Table 2902.1 of IBC 2006 at the following link:
http://publicecodes.citation.com/icod/ibc/2006f2/icod_ibc_2006f2_29_sec002.htm?bu=IC-P-2006-000001&bu2=IC-P-2006-000019

You should become familiar with Table 2902.1 of IBC 2006. It will help you pass the ARE CDS division and help you in your real architectural practice.

90. Answer: a, b, and d
When an architect is doing initial project research, the three most important issues that she needs to consider are:
- zoning
- occupancy
- if the building will be sprinklered

Exit width, plumbing fixtures, and HVAC equipment are less important and can be handled later.

91. Answer: c
The dispute should proceed directly to mediation. Any disputes related to **hazardous** materials should proceed directly to mediation. They are NOT referred to the IDM or the architect.

See comment of Article 10.3.2 of the FREE PDF file of commentary for AIA document A201–2007, General Conditions of the Contract for Construction, at the following link:
http://www.aia.org/contractdocs/aiab081438

92. Answer: d
Coordination of mechanical and electrical ceiling plans is a basic service for the architect, NOT an additional service.

The following are additional service:
- As-constructed record drawings
- Commissioning
- Presentations in a public hearing to defend the owner's choice of water fountains and sculpture

See Article 4.1 of the FREE PDF file of the commentary for B101–2007 (Former B141–1997), Standard Form of Agreement between Owner and Architect (RAIC Document 6) at the following link:
http://www.aia.org/contractdocs/aiab081438

93. Answer: b
A form of strict, secondary liability that arises under the common law doctrine of agency, the responsibility of the superior for the acts of his subordinate is known as **vicarious liability**. Employee liability, parental liability, and principals liability are some examples of vicarious liability.

See Introduction section of the FREE PDF file of commentary for AIA document A201–2007, General Conditions of the Contract for Construction, at the following link:
http://www.aia.org/contractdocs/aiab081438

AND
the definition of vicarious liability at the following link:
http://en.wikipedia.org/wiki/Vicarious_liability

94. Answer: d
All the following can affect a project's construction cost:
- Project location
- Schedule
- Type of construction
- LEED certification

95. Answer: c
Storefront is likely to be shop fabricated and installed at the site. An architect typically reviews shop drawings for storefront.

Plumbing fixtures and HVAC equipment are mass-produced manufactured items.

Slab is placed at the site.

96. Answer: d

Please note we are looking for the incorrect statement as the correct answer.

The following statement regarding construction schedule is incorrect and is therefore the correct answer:

- The **shortest** path is the critical path in CPM.

The following statements regarding construction schedule are correct and are therefore the incorrect answers:

- The contractor is responsible for the construction schedule.
- The architect sets up the criteria for the construction schedule.
- CPM schedule shows the duration, sequences, and relationship between activities.

97. Answer: a

The architect is responsible to the owner for all his work and his consultants' work, including the electrical systems design. The owner has a contract with the architect, NOT his consultants. However, the architect's consultants are responsible to the architect for their portion of the work.

98. Answer: a, b, d, and e

When selecting switchgear, an architect and his electrical engineer should consider the following four most important factors:

- Dimension
- Voltage
- Clearance
- Available space

The following two factors are less important:

- Noise characteristics of the equipment
- Color

99. Answer: c, d, and e

The following are acceptable ways to revise drawings and specifications during construction:

- Change orders
- Written orders for a minor change of the Work
- Construction Change directives

Addenda are ways to revise contract documents, including drawings and specifications before the bid is due. A punch list is a correction items prepared by the contractor for items he has not finished or to be corrected per the contract documents, including drawings and specifications, NOT a way to revise drawings and specifications during construction.

A notice is NOT an official AIA document.

100. Answer: e

Per C401–2007 (Former C141–1997), Standard Form of Agreement Between Architect and Consultant, the consultant shall maintain general liability, automobile liability, workers' compensation, and professional liability insurance for the duration of the agreement.

This is one of changes from the previous version of the form, C141–1997, Standard Form of Agreement Between Architect and Consultant.

One way to remember this:

An architect's contract and insurance requirements will "pass down" to her consultants, i.e., consultants need to carry the same kinds of insurance as the architect.

B. Mock Exam Solution: CDS Graphic vignette section
1. Step-by-step description for a passing solution using the official NCARB CDS practice program

Use the NCARB Practice Program to draw the section:

1) Draw the **grade line**:

Draw > Grade Line

This is the most important step. Your solution will fail if you miss this step. Make sure you leave enough space to draw both the first and second floor sections. Otherwise, you may have to spend more time to move all the elements that you draw later to make room for the second floor section.

Note: Some people prefer to draw the grade line near the section cut line, and overlap the section with the floor plans. This way, they can see the relationship between the section and the floor plans right there without zoom in and out. I personally find this approach confusing and think it makes the drawings too hard to read. Of course, you can have your own preference. Ultimately, it is your choice and work habit.

2) Draw the **exterior walls and bearing walls**:

With the help of the graphic scale on the plans, you can manually scale the horizontal dimensions from the floor plans to locate the section elements.

For this exercise, you can use **sketch > rectangle** to establish the horizontal locations of the walls based on the horizontal dimensions you measured manually per the graphic scale on plans. You can delete the temporary rectangles to avoid too many overlapping elements after you locate the walls.

In the real exam, you can use **sketch > line** to project the exact location of the walls from the plans to the section.

Draw > Exterior/Bearing Wall

You can draw the walls higher than their actual heights for now, and adjust the heights later.

*Note: The left side of your section must align with the line noted as **"EXTERIOR FACE OF SECTION"** on the plans. Otherwise, your solution will automatically fail.*

3) Draw the **slab on grade**:

Draw > Slab on Grade (Figure 4.1)

*Note: The slab is NOT continuous, and it stops at the interior bearing wall, i.e., interior bearing wall sits directly on the foundation below and divides the slab into two segments on this section. You can use the **Zoom** command to zoom in and adjust or move the element you draw to a more accurate location.*

For example, if we notice the slab on grade is too far above the grade line, we zoom in to adjust it to right above the grade line:

Zoom > click on the rectangle area where you want to zoom in.

Use **Move, Adjust** to move the slab on grade to the desired location, and then click on the location to place the slab on grade on the new location. (Figure 4.2)

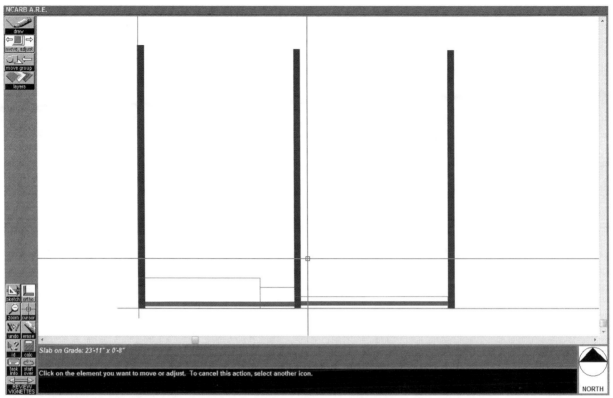

Figure 4.1 Draw Grade Line, Exterior/Bearing Walls, and Slab on Grade

4) Draw **foundation with footing under the exterior walls**

Zoom in to the area you want to work on.

Draw > Foundation with Footing (Figure 4.3)

Per the program of this vignette, the frost depth is 4'-6" (1372), so make sure the <u>bottom</u> of the foundation is placed <u>exactly</u> at 4'-6" (1372) below the grade line to match the frost depth. The foundation depth automatically shows on the lower left hand corner of the computer screen when you draw the foundation.

Use **Move, Adjust** to move exterior foundation with footing to align with the exterior wall (Figure 4.4).

Use the same procedure to draw the other exterior foundation with footing.

Note:
1. You should always place the <u>bottom</u> instead of the <u>top</u> of the exterior footings at frost line. The reasons are:

Most soils expand and heave when they freeze.

*The **frost line** is the depth at which the moisture present in the soil is expected to freeze. Once the bottom of the footing is placed at the frost line the ground will act as a barrier to insulate the soil below the footing from freezing in the winter, and thus prevent the soils below the frost line from expanding and heaving.*

If you place the <u>top</u> of the footing at the frost line, it will still work, but it will cost MORE money, and it is not as good as placing the <u>bottom</u> of the footing at the frost line. You are supposed to draw the <u>minimum</u> depth for the footings for the CDS graphic vignette.

2. *You should always align the <u>top</u> of the foundation with footings with the grade line (the <u>bottom</u> of the slab), NOT the <u>top</u> of the slab. The reason is:*

If you align the <u>top</u> of the foundation with footings with the <u>top</u> of the slab, and you draw a 4'-6" deep foundation with footing using the NCARB graphic software, your bottom of the footing will be a few inches higher than the frost line because of the slab thickness, and your solution will probably fail.

5) Use the same procedure to draw the **foundation for the interior bearing wall**, except the top of the foundation should be placed <u>right below the bottom of the slab</u>, NOT close to the frost depth. The interior bearing wall is typically thicker, and you can confirm it is an interior bearing wall by turning on only the layers for the 1st floor, or only the layer for the 2nd floor to see the related structural joists.

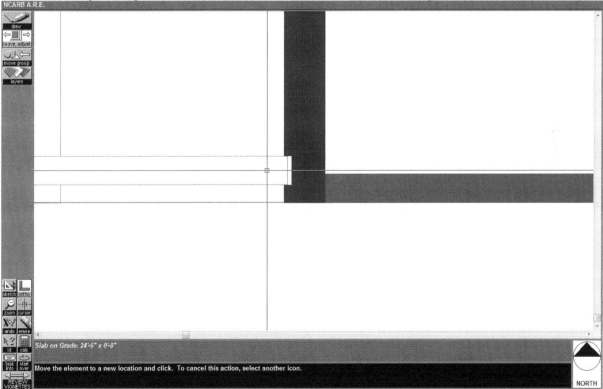

Figure 4.2   Zoom in and use **Move, Adjust** to move the slab on grade to an accurate location.

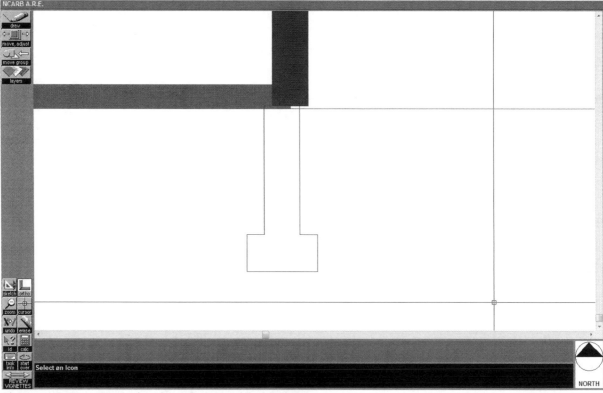

Figure 4.3 Draw exterior foundation with footing.

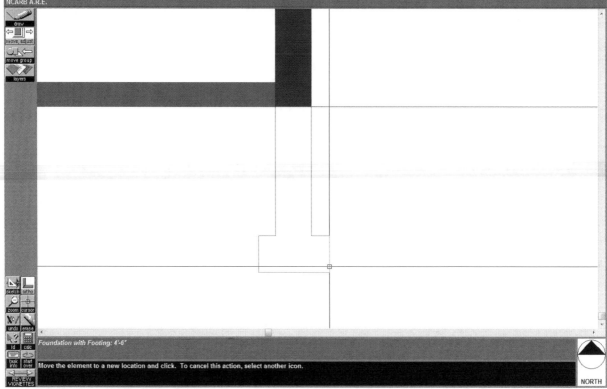

Figure 4.4   Use **Move, Adjust** to move exterior foundation with footing to align with the exterior wall.

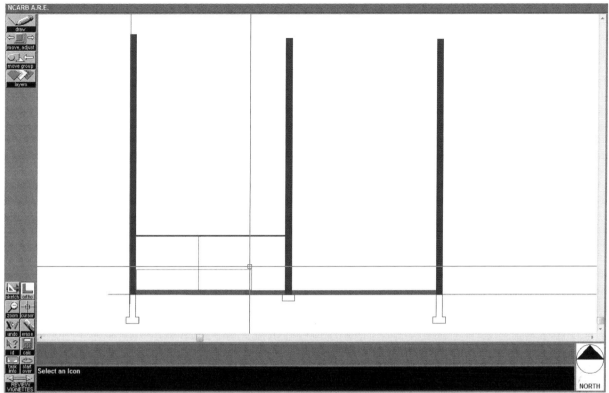

Figure 4.5   Use **Sketch > Rectangle** to establish the ceiling height, and use **Draw > Finished Ceiling** to draw the ceiling.

6) Use **Sketch > Rectangle** to draw an 8'-6" (2590) high rectangle to establish the ceiling height for the office and the corridor.

Note: The ceiling height is measured from the <u>top</u> of the slab on grade to the <u>bottom</u> of the ceiling.

7) Use **Draw > Finished Ceiling** to draw the ceiling for the office and the corridor (Figure 4.5).

8) Use **Sketch > Rectangle** to draw an 8" (203) high rectangle to establish the **clearance for light fixtures**.

9) Since all the joists for the 1st floor space (except the double-height multi-purpose room) are the same size, we can use the **deepest** duct for the first floor to establish the **clearance for the ducts.** The **deepest** duct for the first floor is 32x20 (813x508), and it is 20" (508) deep. Use **Sketch > Rectangle** to draw a 20" (508) high rectangle to establish the **clearance for the ducts.**

10) For the office and corridor: Since the joists are cut by the section cut line, they should be shown as sectional view. Since the 1st floor plan notes 32" (813) joists @ 3'-0" (36" or 914) O.C., we use **Draw > Joists with Deck (Section) > Depth 32"** (813) **> Spacing 36"** (914) to draw the joists for the laboratory space.

11) Use **Move Group** to move the entire joist and align the bottom of the joists with the top of the duct clearance line, and align the first joist and last joist with the ones on the plan (Figure 4.6).

12) Use **Zoom** and **Move, Adjust** to extend both ends of the deck to 4" into the exterior wall and the interior bearing wall.

*Note: You can use **Move Group** to move the entire joist run, and **Move, Adjust** to adjust the locations of the joists within the joist run. Make sure the locations of your joists on the section MATCH the locations of the joists on the floor plans. Otherwise, you may fail this vignette.*

13) Use **Draw > Duct > Width 32"** (813) > **Depth 20"** (508) to draw the 32x20 (813x508) duct at the section cut line, align it with floor plan, and align the top of the duct to the bottom of the joists (Figure 4.7). You can see the elevation of the 12x12 duct adjacent to the cut line, but NCARB Practice Program does NOT have the function of drawing duct elevation, so you do NOT have to show the duct elevation for this practice.
14) Use **Draw > Interior Fire-Rated Partition** to draw the wall between the office and the corridor. The rated wall should be extended to the bottom of deck
15) The storage room wall is NOT fire-rated, and should be extended to 6" (152) above the finish ceiling. Use **Draw > Interior Partition** to draw it.
16) Repeat steps 7) to 15) to draw the remaining rooms for the 1st and 2nd floors (Figure 4.8).
17) Use **Sketch > Rectangle** to draw a 3'-0" (914) high rectangle to establish the top of parapet height.
18) Use **Move, Adjust** to adjust the top of parapet to 3'-0" (914) above the highest adjacent roof, as required by the program (Figure 4.9).
19) Use **Sketch > Hide Sketches Elements** to show the completed section without the sketch elements (Figure 4.10).
20) Use the Sketch tools or ID tools to double check all dimensions, heights, and clearances.

You should practice afore-mentioned steps 1) to 19) a few times to become familiar with the process, understand the pitfalls, and master the NCARB CDS practice program. You should be able to finish the graphic vignette within the required 1 hour testing time.

2. Notes on graphic vignette traps
Pay attention to the following graphic **vignette** traps:
- The ceiling height for the 2nd floor is 9'-0" (2743) instead of 8'-6" (2590)
- Do NOT forget the 8" (203) clearance height needed for light fixtures.
- All corridor walls are fire-rated, and should be extended to the bottom of deck. Use **Draw > Interior Fire-Rated Partition** to draw them.
- Align the section elements (ducts, joists and interior partitions) with those on the floor plans.
- Make sure you draw the ducts and joists at the <u>correct</u> sizes (and <u>correct</u> spacing for joists) per the plans.
- Draw room heights, parapet heights, clearance for ducts and light fixtures per the **<u>minimum</u>** dimensions as required by the program.
- The multi-purpose room is a **double height space** and has its own joist depth. The duct at the section cut line is 24x12 (610x305). However, the **<u>deepest</u>** duct for the multi-

purpose room is 32x24 (813x610) with a depth of 24" (610). Since **ALL** ducts are placed below the joists, you should use 24" (610) as the duct clearance depth for the multi-purpose room.

- The joists at the multi-purpose room are 36" deep at an elevation view instead of a section view. Use **Draw > Joists with Deck (Elevation) > Depth 36"** to draw them.
- In the real exam, use **Layers** to turn on only the 2nd floor layers. Use **Sketch > Line** to project the locations of the first and last joists for the laboratory space.

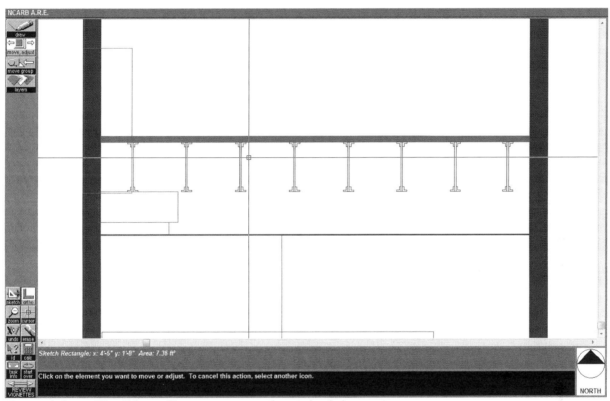

Figure 4.6   Use **Move Group** to move the entire joist.

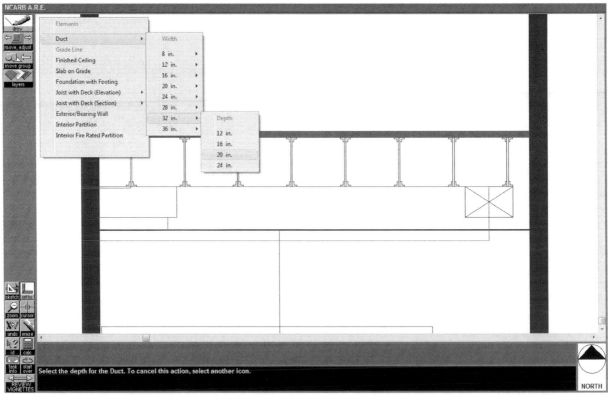

Figure 4.7   Use **Draw > Duct > Width 32"** (813) > **Depth 20"** (508) to draw the 32x20 (813x508) duct at section cut line.

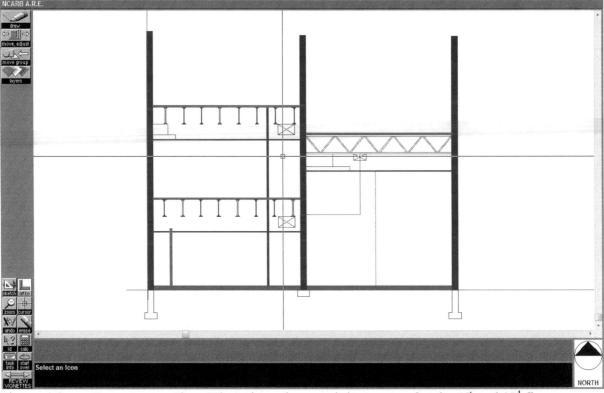

Figure 4.8   Repeat steps 7) to 15) to draw the remaining rooms for the 1st and 2nd floors.

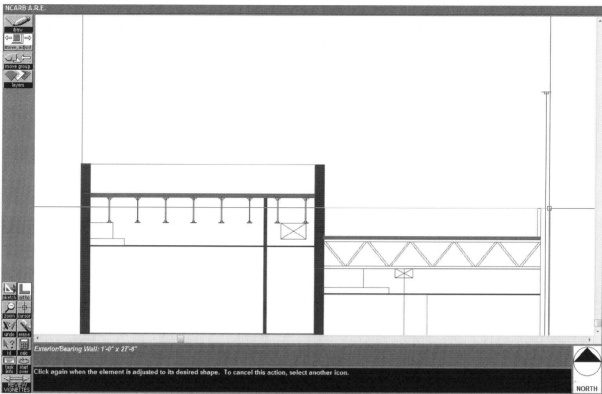

Figure 4.9   Use **Move, Adjust** to adjust the top of parapet to 3'-0" (914) above the highest adjacent roof, as required by the program.

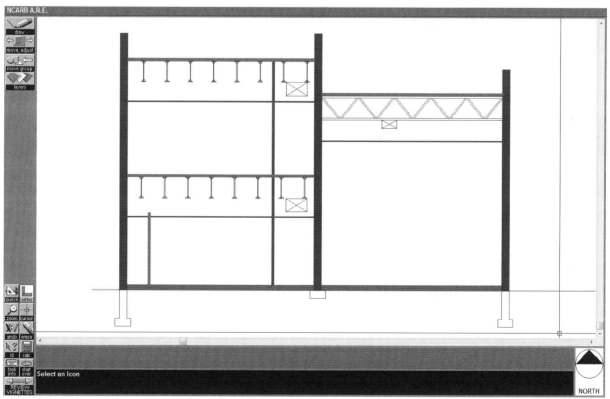

Figure 4.10   The passing solution with the sketch elements hidden

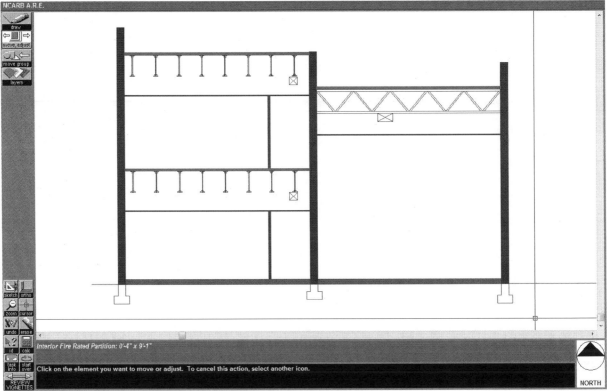

Figure 4.11   A failing solution for CDS graphic vignette

3. **A summary of the critical dimensions**

For your convenience, the following is a summary of the critical dimensions:

Frost depth (<u>Bottom</u> of the exterior footings)
4'-6" (1371) below the grade line (or the <u>bottom</u>, NOT the top of the slab)

Multiple-purpose room (Double Height Space)
Ceiling – 18'-0" (5486)
Lights – 8" (203)
Worst case duct depth – 24" (610)
Joists – 36" @ 5'-0" (914 @ 1524)
Parapet – 36" (914)
Duct size at section cutline – 24x12 (610 x 305)

The remaining First Floor
Ceiling – 8'-6" (2590)
Lights –8" (203)
Worst case duct depth – 20" (508)
Joists – 32" @ 3'-0" O.C. (813 @ 914)
Duct size at section cutline – 32x20 (813 x 508)

Second Floor
Ceiling – 9'-0" (2743)
Lights – 8" (203)
Worst case duct depth – 20" (508)
Joists – 32" @ 3'-0" O.C. (813 @ 914)
Parapet – 36" (914)
Duct size at section cutline – 32x20 (813 x 508)

4. A failing solution and lessons learned

Figure 4.11 shows a failing solution. This solution has the following errors:

1) The footings for exterior walls are NOT deep enough, and the bottoms of the footing are above frost line.
2) The footing for the interior bearing wall is too deep. Its top should be placed right below the slap.
3) The top of the parapet for the interior bearing wall is NOT 3'-0" (914) above the highest adjacent roof, as required by the program.
4) The ducts at the corridors are too small.
5) The interior partition for the storage room is missing.
6) The fire-rated corridor walls are NOT extended to the bottom of decks.

Appendixes

A. List of Figures

B. Official reference materials suggested by NCARB

1. General NCARB reference materials for ARE:

Per NCARB, all candidates should become familiar with the latest version of the following codes:

International Code Council, Inc. (ICC, 2006)
International Building Code
International Mechanical Code
International Plumbing Code

National Fire Protection Association (NFPA)
Life Safety Code (NFPA 101)
National Electrical Code (NFPA 70)

National Research Council of Canada
National Building Code of Canada
National Plumbing Code of Canada
National Fire Code of Canada

American Institute of Architects
AIA Documents - 2007

Candidates should be familiar with the Standard on Accessible and Usable Buildings and Facilities (ICC/ANSI A117.1-98)

2. Official NCARB reference materials for the Construction Documents & Services (CDS) division:

American Institute of Architects, AIA Documents (especially **A101**, **A201** and **B101**), American Institute of Architects, Washington, DC, latest edition. Starting July 2010, Construction Documents & Services will reference **2007 AIA Documents**. The 14th edition of *The Architect's Handbook of Professional Practice* has a CD containing the PDF files of the sample 2007 AIA contract documents.

See following link for synopses of 115 AIA documents by project type or delivery method in eight document families:
http://www.aia.org/contractdocs/referencematerial/index.htm

Ching, Francis and Cassandra Adams. *Building Construction Illustrated.* Wiley, latest edition. The illustrations are great! Ching has a great talent for simplifying complicated issues and make them very easy to understand.

Committee of Canadian Architectural Councils and The Royal Architectural Institute of Canada. *Canadian Handbook of Practice for Architects*. Committee of Canadian Architectural Councils and The Royal Architectural Institute of Canada, latest edition

Construction Specifications Institute. *The Project Resource Manual: CSI Manual of Practice*. Construction Specifications Institute, Alexandria, VA, latest edition. A valuable resource for the organization, preparation, use, and interpretation of construction documents encompassing the entire life cycle of a facility.

Demkin, Joseph A., AIA, Executive Editor. *The Architect's Handbook of Professional Practice* (AHPP). The American Institute of Architects & Wiley, latest edition. A comprehensive book covers all aspect of architectural practice, including 2 CDs containing the sample AIA contract documents.

National Council of Architectural Registration Boards (NCARB). *Rules of Conduct*. National Council of Architectural Registration Boards, latest edition. (Available as a FREE PDF file at http://www.ncarb.org/)

Ramsey, Charles George, and John Ray Hoke Jr. *Architectural Graphic Standards*. The American Institute of Architects & Wiley, latest edition. It is organized roughly per the CSI MasterFormat divisions, including general planning and design data, site work, concrete, masonry, metals, wood, thermal and moisture protection, doors and windows, finishes, specialties, equipment, furnishings, special construction, conveying systems, mechanical, electrical, sports, energy, history preservation, etc.

C. Other reference materials

Access Board, *ADAAG Manual: A Guide to the American with Disabilities Accessibility Guidelines*. East Providence, RI: BNI Building News. ADA Standards for Accessible Design are available as FREE PDF files at

http://www.ada.gov/

AND
http://www.access-board.gov/adaag/html/figures/index.html

American Institute of Steel Construction. *Steel Construction Manual, 13th Edition*. American Institute of Steel Construction. latest edition. This is a very expensive but also very valuable book for both architects and structural engineers. For architects, you can use this book to look up the sectional dimensions for columns and beams and their structural weight per linear foot, and then you can find out if your columns will fit inside an interior partition or find out the height of the beams. Then you can subtract the beam height, mechanical duct height, and T-bar ceiling height and floor thickness to calculate the actual ceiling heights for space under a mezzanine, etc.

Chen, Gang. *Building Construction: Project Management, Construction Administration, Drawings, Specs, Detailing Tips, Schedules, Checklists, and Secrets Others Don't Tell You (Architectural Practice Simplified, 2nd edition).* ArchiteG, Inc., A good introduction to the architectural practice and construction documents and service, including discussions of MasterSpec format and specification sections.

Chen, Gang. *LEED GA Exam Guide: A Must-Have for the LEED Green Associate Exam: Comprehensive Study Materials, Sample Questions, Mock Exam, Green Building LEED Certification, and Sustainability.* ArchiteG, Inc., latest edition. A good introduction to green buildings and the LEED building rating system.

Ching, Francis. *Architecture: Form, Space, & Order.* Wiley, latest edition. It is one of the best architectural books that you can have. I still flip through it every now and then. It is a great book for inspiration.

Ching, Francis. Steven R. Winkel, FAIA, PE. *Building Codes Illustrated: A Guide to Understanding the International Building Code.* Wiley, latest edition. A valuable interpretive guide with many useful line drawings. A great timesaver.

Frampton, Kenneth. *Modern Architecture: A Critical History.* Thames and Hudson, London, latest edition. A valuable resource for architectural history.

Jarzombek, Mark M. (Author), Vikramaditya Prakash (Author), Francis D. K. Ching (Editor). *A Global History of Architecture.* Wiley, latest edition. A valuable and comprehensive resource for architectural history with 1000 b & w photos, 50 color photos, and 1500 b & w illustrations. It doesn't limit the topic on a Western perspective, but rather through a global vision.

Trachtenberg, Marvin and Isabelle Hyman. *Architecture: From Pre-history to Post-Modernism.* Prentice Hall, Englewood Cliffs, NJ, latest edition. A valuable and comprehensive resource for architectural history.

D. Definition of Architects and Some Important Information about Architects and the Profession of Architecture

Architects, Except Landscape and Naval

- Nature of the Work
- Training, Other Qualifications, and Advancement
- Employment
- Job Outlook
- Projections Data
- Earnings
- OES Data
- Related Occupations
- Sources of Additional Information

Significant Points

- About 1 in 5 architects are self-employed—more than 2 times the proportion for all occupations.
- Licensing requirements include a professional degree in architecture, at least 3 years of practical work training, and passing all divisions of the Architect Registration Examination.
- Architecture graduates may face competition, especially for jobs in the most prestigious firms.

Nature of the Work

People need places in which to live, work, play, learn, worship, meet, govern, shop, and eat. These places may be private or public; indoors or out; rooms, buildings, or complexes, and architects design them. Architects are licensed professionals trained in the art and science of building design who develop the concepts for structures and turn those concepts into images and plans.

Architects create the overall aesthetic and look of buildings and other structures, but the design of a building involves far more than its appearance. Buildings also must be functional, safe, and economical and must suit the needs of the people who use them. Architects consider all these factors when they design buildings and other structures.

Architects may be involved in all phases of a construction project, from the initial discussion with the client through the entire construction process. Their duties require specific skills—designing, engineering, managing, supervising, and communicating with clients and builders. Architects spend a great deal of time explaining their ideas to clients, construction contractors, and others. Successful architects must be able to communicate their unique vision persuasively.

The architect and client discuss the objectives, requirements, and budget of a project. In some cases, architects provide various pre-design services: conducting feasibility and environmental impact studies, selecting a site, preparing cost analysis and land-use studies, or specifying the requirements the design must meet. For example, they may determine space requirements by researching the numbers and types of potential users of a building. The architect then prepares drawings and a report presenting ideas for the client to review.

After discussing and agreeing on the initial proposal, architects develop final construction plans that show the building's appearance and details for its construction. Accompanying these plans are drawings of the structural system; air-conditioning, heating, and ventilating systems; electrical systems; communications systems; plumbing; and, possibly, site and landscape plans. The plans also specify the building materials and, in some cases, the interior furnishings. In developing designs, architects follow building codes, zoning laws, fire regulations, and other ordinances, such as those requiring easy access by people who are disabled. Computer-aided design and drafting (CADD) and Building Information Modeling (BIM) technology has replaced traditional paper and pencil as the most common method for creating design and construction drawings. Continual revision of plans on the basis of client needs and budget constraints is often necessary.

Architects may also assist clients in obtaining construction bids, selecting contractors, and negotiating construction contracts. As construction proceeds, they may visit building sites to make sure that contractors follow the design, adhere to the schedule, use the specified materials, and meet work quality standards. The job is not complete until all construction is finished, required tests are conducted, and construction costs are paid. Sometimes, architects also provide post-construction services, such as facilities management. They advise on energy efficiency measures, evaluate how well the building design adapts to the needs of occupants, and make necessary improvements.

Often working with engineers, urban planners, interior designers, landscape architects, and other professionals, architects in fact spend a great deal of their time coordinating information from, and the work of, other professionals engaged in the same project.

They design a wide variety of buildings, such as office and apartment buildings, schools, churches, factories, hospitals, houses, and airport terminals. They also design complexes such as urban centers, college campuses, industrial parks, and entire communities.

Architects sometimes specialize in one phase of work. Some specialize in the design of one type of building—for example, hospitals, schools, or housing. Others focus on planning and pre-design services or construction management and do minimal design work.

Work environment. Usually working in a comfortable environment, architects spend most of their time in offices consulting with clients, developing reports and drawings, and working with other architects and engineers. However, they often visit construction sites to review the progress of projects. Although most architects work approximately 40 hours per week, they often have to work nights and weekends to meet deadlines.

Training, Other Qualifications, and Advancement

There are three main steps in becoming an architect. First is the attainment of a professional degree in architecture. Second is work experience through an internship, and third is licensure through the passing of the Architect Registration Exam.

Education and training. In most States, the professional degree in architecture must be from one of the 114 schools of architecture that have degree programs accredited by the National Architectural Accrediting Board. However, State architectural registration boards set their own standards, so graduation from a non-accredited program may meet the educational requirement for licensing in a few States.

Three types of professional degrees in architecture are available: a 5-year bachelor's degree, which is most common and is intended for students with no previous architectural training; a 2-year master's degree for students with an undergraduate degree in architecture or a related area; and a 3- or 4-year master's degree for students with a degree in another discipline.

The choice of degree depends on preference and educational background. Prospective architecture students should consider the options before committing to a program. For example, although the 5-year bachelor of architecture offers the fastest route to the professional degree, courses are specialized, and if the student does not complete the program, transferring to a program in another discipline may be difficult. A typical program includes courses in architectural history and theory, building design with an emphasis on CADD, structures, technology, construction methods, professional practice, math, physical sciences, and liberal arts. Central to most architectural programs is the design studio, where students apply the skills and concepts learned in the classroom, creating drawings and three-dimensional models of their designs.

Many schools of architecture also offer post-professional degrees for those who already have a bachelor's or master's degree in architecture or other areas. Although graduate education beyond the professional degree is not required for practicing architects, it may be required for research, teaching, and certain specialties.

All State architectural registration boards require architecture graduates to complete a training period—usually at least 3 years—before they may sit for the licensing exam. Every State, with the exception of Arizona, has adopted the training standards established by the Intern Development Program, a branch of the American Institute of Architects and the National Council of Architectural Registration Boards (NCARB). These standards stipulate broad training under the supervision of a licensed architect. Most new graduates complete their training period by working as interns at architectural firms. Some States allow a portion of the training to occur in the offices of related professionals, such as engineers or general contractors. Architecture students who complete internships while still in school can count some of that time toward the 3-year training period.

Interns in architectural firms may assist in the design of one part of a project, help prepare architectural documents or drawings, build models, or prepare construction drawings on

CADD. Interns also may research building codes and materials or write specifications for building materials, installation criteria, the quality of finishes, and other, related details.

Licensure. All States and the District of Columbia require individuals to be licensed (registered) before they may call themselves architects and contract to provide architectural services. During the time between graduation and becoming licensed, architecture school graduates generally work in the field under the supervision of a licensed architect who takes legal responsibility for all work. Licensing requirements include a professional degree in architecture, a period of practical training or internship, and a passing score on all divisions of the Architect Registration Examination. The examination is broken into nine divisions consisting of either multiple choice or graphical questions. The eligibility period for completion of all divisions of the exam varies by State.

Most States also require some form of continuing education to maintain a license, and many others are expected to adopt mandatory continuing education. Requirements vary by State but usually involve the completion of a certain number of credits annually or biennially through workshops, formal university classes, conferences, self-study courses, or other sources.

Other qualifications. Architects must be able to communicate their ideas visually to their clients. Artistic and drawing ability is helpful, but not essential, to such communication. More important are a visual orientation and the ability to understand spatial relationships. Other important qualities for anyone interested in becoming an architect are creativity and the ability to work independently and as part of a team. Computer skills are also required for writing specifications, for 2- and 3- dimensional drafting using CADD programs, and for financial management.

Certification and advancement. A growing number of architects voluntarily seek certification by the National Council of Architectural Registration Boards. Certification is awarded after independent verification of the candidate's educational transcripts, employment record, and professional references. Certification can make it easier to become licensed across States. In fact, it is the primary requirement for reciprocity of licensing among State Boards that are NCARB members. In 2007, approximately one-third of all licensed architects had this certification.

After becoming licensed and gaining experience, architects take on increasingly responsible duties, eventually managing entire projects. In large firms, architects may advance to supervisory or managerial positions. Some architects become partners in established firms, while others set up their own practices. Some graduates with degrees in architecture also enter related fields, such as graphic, interior, or industrial design; urban planning; real estate development; civil engineering; and construction management.

Employment

Architects held about 132,000 jobs in 2006. Approximately 7 out of 10 jobs were in the architectural, engineering, and related services industry—mostly in architectural firms with fewer than five workers. A small number worked for residential and nonresidential building construction firms and for government agencies responsible for housing, community planning, or construction of government buildings, such as the U.S. Departments of Defense and Interior, and the General Services Administration. About 1 in 5 architects are self-employed.

Job Outlook

Employment of architects is expected to grow faster than the average for all occupations through 2016. Keen competition is expected for positions at the most prestigious firms, and opportunities will be best for those architects who are able to distinguish themselves with their creativity.

Employment change. Employment of architects is expected to grow by 18 percent between 2006 and 2016, which is <u>faster than the average</u> for all occupations. Employment of architects is strongly tied to the activity of the construction industry. Strong growth is expected to come from nonresidential construction as demand for commercial space increases. Residential construction, buoyed by low interest rates, is also expected to grow as more people become homeowners. If interest rates rise significantly, home building may fall off, but residential construction makes up only a small part of architects' work.

Current demographic trends also support an increase in demand for architects. As the population of Sunbelt States continues to grow, the people living there will need new places to live and work. As the population continues to live longer and baby-boomers begin to retire, there will be a need for more healthcare facilities, nursing homes, and retirement communities. In education, buildings at all levels are getting older and class sizes are getting larger. This will require many school districts and universities to build new facilities and renovate existing ones.

In recent years, some architecture firms have outsourced the drafting of construction documents and basic design for large-scale commercial and residential projects to architecture firms overseas. This trend is expected to continue and may have a negative impact on employment growth for lower level architects and interns who would normally gain experience by producing these drawings.

Job prospects. Besides employment growth, additional job openings will arise from the need to replace the many architects who are nearing retirement, and others who transfer to other occupations or stop working for other reasons. Internship opportunities for new architectural students are expected to be good over the next decade, but more students are graduating with architectural degrees and some competition for entry-level jobs can be anticipated. Competition will be especially keen for jobs at the most prestigious architectural firms as prospective architects try to build their reputation. Prospective architects who have had internships while in school will have an advantage in obtaining intern positions after graduation. Opportunities will be best for those architects that are able to distinguish themselves from others with their creativity.

Prospects will also be favorable for architects with knowledge of "green" design. Green design, also known as sustainable design, emphasizes energy efficiency, renewable resources such as energy and water, waste reduction, and environmentally friendly design, specifications, and materials. Rising energy costs and increased concern about the environment has led to many new buildings being built green.

Some types of construction are sensitive to cyclical changes in the economy. Architects seeking design projects for office and retail construction will face especially strong competition for jobs or clients during recessions, and layoffs may ensue in less successful firms. Those involved in the design of institutional buildings, such as schools, hospitals, nursing homes, and correctional facilities, will be less affected by fluctuations in the economy. Residential construction makes up a small portion of work for architects, so major changes in the housing market would not be as significant as fluctuations in the nonresidential market.

Despite good overall job opportunities, some architects may not fare as well as others. The profession is geographically sensitive, and some parts of the Nation may have fewer new building projects. Also, many firms specialize in specific buildings, such as hospitals or office towers, and demand for these buildings may vary by region. Architects may find it increasingly necessary to gain reciprocity in order to compete for the best jobs and projects in other States.

Projections Data

Projections data from the National Employment Matrix

Occupational title	SOC Code	Employment, 2006	Projected employment, 2016	Change, 2006-16		Detailed statistics
				Number	Percent	
Architects, except landscape and naval	17-1011	132,000	155,000	23,000	18	PDF zipped XLS
NOTE: Data in this table are rounded. See the discussion of the employment projections table in the *Handbook* introductory chapter on *Occupational Information Included in the Handbook*.						

Earnings

Median annual earnings of wage-and-salary architects were $64,150 in May 2006. The middle 50 percent earned between $49,780 and $83,450. The lowest 10 percent earned less than $39,420, and the highest 10 percent earned more than $104,970. Those just starting their internships can expect to earn considerably less.

Earnings of partners in established architectural firms may fluctuate because of changing business conditions. Some architects may have difficulty establishing their own practices and may go through a period when their expenses are greater than their income, requiring substantial financial resources.

Many firms pay tuition and fees toward continuing education requirements for their employees.

For the latest wage information:
The above wage data are from the Occupational Employment Statistics (OES) survey program, unless otherwise noted. For the latest National, State, and local earnings data, visit the following pages:

Architects, except landscape and naval

Related Occupations

Architects design buildings and related structures. Construction managers, like architects, also plan and coordinate activities concerned with the construction and maintenance of buildings and facilities. Others who engage in similar work are landscape architects, civil engineers, urban and regional planners, and designers, including interior designers, commercial and industrial designers, and graphic designers.

Sources of Additional Information

Disclaimer:
Links to non-BLS Internet sites are provided for your convenience and do not constitute an endorsement.

Information about education and careers in architecture can be obtained from:

- The American Institute of Architects, 1735 New York Ave. NW., Washington, DC 20006. Internet: http://www.aia.org
- Intern Development Program, National Council of Architectural Registration Boards, Suite 1100K, 1801 K St. NW., Washington, D.C. 20006. Internet: http://www.ncarb.org OOH ONET Codes 17-1011.00"

Quoted from: Bureau of Labor Statistics, U.S. Department of Labor, Occupational Outlook Handbook, 2008-09 Edition, Architects, Except Landscape and Naval, on the Internet at **http://www.bls.gov/oco/ocos038.htm** (visited November 30, 2008).
Last Modified Date: December 18, 2007

Note: Please check the website above for the latest information.

E. AIA Compensation Survey

Every 3 years, AIA publishes a Compensation Survey for various positions at architectural firms across the country. It is a good idea to find out the salary before you make the final decision to become an architect. If you are already an architect, it is also a good idea to determine if you are underpaid or overpaid.

See following link for some sample pages for the 2008 AIA Compensation Survey:

http://www.aia.org/aiaucmp/groups/ek_public/documents/pdf/aiap072881.pdf

F. So … You would Like to Study Architecture

To study architecture, you need to learn how to draft, how to understand and organize spaces and the interactions between interior and exterior spaces, how to do design, and how to communicate effectively. You also need to understand the history of architecture.

As an architect, a leader for a team of various design professionals, you not only need to know architecture, but also need to understand enough of your consultants' work to be able to coordinate them. Your consultants include soils and civil engineers, landscape architects, structural, electrical, mechanical, and plumbing engineers, interior designers, sign consultants, etc.

There are two major career paths for you in architecture: practice as an architect or teach in colleges or universities. The earlier you determine which path you are going to take, the more likely you will be successful at an early age. Some famous and well-respected architects, like my USC alumnus Frank Gehry, have combined the two paths successfully. They teach at the universities and have their own architectural practice. Even as a college or university professor, people respect you more if you have actual working experience and have some built projects. If you only teach in colleges or universities but have no actual working experience and have no built projects, people will consider you as a "paper" architect, and they are not likely to take you seriously, because they will think you probably do not know how to put a real building together.

In the U.S., if you want to practice architecture, you need to obtain an architect's license. It requires a combination of passing scores on the Architectural Registration Exam (ARE) and 8 years of education and/or qualified working experience, including at least 1 year of working experience in the U.S. Your working experience needs to be under the supervision of a licensed architect to be counted as qualified working experience for your architect's license.

If you work for a landscape architect or civil engineer or structural engineer, some states' architectural licensing boards will count your experience at a discounted rate for the qualification of your architect's license. For example, 2 years of experience working for a civil engineer may be counted as 1 year of qualified experience for your architect's license. You need to contact your state's architectural licensing board for specific licensing requirements for your state.

If you want to teach in colleges or universities, you probably want to obtain a master's degree or a Ph.D. It is not very common for people in the architectural field to have a Ph.D. One reason is that there are few Ph.D. programs for architecture. Another reason is that architecture is considered a profession and requires a license. Many people think an architect's license is more important than a Ph.D. degree. In many states, you need to have an architect's license to even use the title "architect," or the terms "architectural" or "architecture" to advertise your service. You cannot call yourself an architect if you do not have an architect's license, even if you have a Ph.D. in architecture. Violation of these rules brings punishment.

To become a tenured professor, you need to have a certain number of publications and pass the evaluation for the tenure position. Publications are very important for tenure track positions. Some people say for the tenured track positions in universities and colleges, it is "publish or perish."

The American Institute of Architects (AIA) is the national organization for the architectural profession. Membership is voluntary. There are different levels of AIA membership. Only licensed architects can be (full) AIA members. If you are an architectural student or an intern but not a licensed architect yet, you can join as an associate AIA member. Contact AIA for detailed information.

The National Council of Architectural Registration Boards (NCARB) is a nonprofit federation of architectural licensing boards. It has some very useful programs, such as IDP, to assist you in obtaining your architect's license. Contact NCARB for detailed information.

Back Page Promotion

You may be interested in some other books written by Gang Chen:

A. **ARE Mock Exam series. See the following link:**
    http://www.GreenExamEducation.com

B. **LEED Exam Guides series. See the following link:**
    http://www.GreenExamEducation.com

C. ***Building Construction:*** *Project Management, Construction Administration, Drawings, Specs, Detailing Tips, Schedules, Checklists, and Secrets Others Don't Tell You (Architectural Practice Simplified, 2nd edition)*
    http://www.ArchiteG.com

D. ***Planting Design Illustrated***
    http://outskirtspress.com/agent.php?key=11011&page=GangChen

ARE Mock Exam Series

Published ARE books (One Mock Exam book for each ARE division, plus California Supplemental Mock Exam):
Programming, Planning & Practice (PPP) ARE Mock Exam (Architect Registration Exam): ARE Overview, Exam Prep Tips, Multiple-Choice Questions and Graphic Vignettes, Solutions and Explanations. **ISBN-13:** 9781612650067

Site Planning & Design ARE Mock Exam (SPD of Architect Registration Exam): ARE Overview, Exam Prep Tips, Multiple-Choice Questions and Graphic Vignettes, Solutions and Explanations. **ISBN-13:** 9781612650111

Building Design and Construction Systems (BDCS) ARE Mock Exam (Architect Registration Exam): ARE Overview, Exam Prep Tips, Multiple-Choice Questions and Graphic Vignettes, Solutions and Explanations. **ISBN-13:** 9781612650029

Schematic Design (SD) ARE Mock Exam (Architect Registration Exam): ARE Overview, Exam Prep Tips, Graphic Vignettes, Solutions and Explanations
ISBN: 9781612650050

Structural Systems ARE Mock Exam (SS of Architect Registration Exam): ARE Overview, Exam Prep Tips, Multiple-Choice Questions and Graphic Vignettes, Solutions and Explanations. **ISBN**: 9781612650012

Building Systems (BS) ARE Mock Exam (Architect Registration Exam): ARE Overview, Exam Prep Tips, Multiple-Choice Questions and Graphic Vignettes, Solutions and Explanations. **ISBN-13**: 9781612650036

Construction Documents and Service (CDS) Are Mock Exam (Architect Registration Exam): ARE Overview, Exam Prep Tips, Multiple-Choice Questions and Graphic Vignettes, Solutions and Explanations. **ISBN-13:** 9781612650005

Mock California Supplemental Exam (CSE of Architect Registration Exam): CSE Overview, Exam Prep Tips, General Section and Project Scenario Section, Questions, Solutions and Explanations. **ISBN**: 9781612650159

Upcoming ARE books:
Other books in the ARE Mock Exam Series are being produced. Our goal is to produce one mock exam book PLUS one guidebook for each of the ARE exam divisions.

See the following link for the latest information:
http://www.GreenExamEducation.com

LEED Exam Guides series*: Comprehensive Study Materials, Sample Questions, Mock Exam, Building LEED Certification and Going Green

LEED (Leadership in Energy and Environmental Design) is the most important trend of development, and it is revolutionizing the construction industry. It has gained tremendous momentum and has a profound impact on our environment.

From LEED Exam Guides series, you will learn how to

1. Pass the LEED Green Associate Exam and various LEED AP + exams (each book will help you with a specific LEED exam).

2. Register and certify a building for LEED certification.

3. Understand the intent for each LEED prerequisite and credit.

4. Calculate points for a LEED credit.

5. Identify the responsible party for each prerequisite and credit.

6. Earn extra credit (exemplary performance) for LEED.

7. Implement the local codes and building standards for prerequisites and credit.

8. Receive points for categories not yet clearly defined by USGBC.

There is currently NO official book on the LEED Green Associate Exam, and most of the existing books on LEED and LEED AP are too expensive and too complicated to be practical and helpful. The pocket guides in LEED Exam Guides series fill in the blanks, demystify LEED, and uncover the tips, codes, and jargon for LEED as well as the true meaning of "going green." They will set up a solid foundation and fundamental framework of LEED for you. Each book in the LEED Exam Guides series covers every aspect of one or more specific LEED rating system(s) in plain and concise language and makes this information understandable to all people.

These pocket guides are small and easy to carry around. You can read them whenever you have a few extra minutes. They are indispensable books for all people—administrators; developers; contractors; architects; landscape architects; civil, mechanical, electrical, and plumbing engineers; interns; drafters; designers; and other design professionals.

Why is the LEED Exam Guides series needed?

A number of books are available that you can use to prepare for the LEED exams:

1. *USGBC Reference Guides.* You need to select the correct version of the *Reference Guide* for your exam.

   The *USGBC Reference Guides* are comprehensive, but they give too much information. For example, *The LEED 2009 Reference Guide for Green Building Design and Construction (BD&C)* has about 700 oversized pages. Many of the calculations in the books are too detailed for the exam. They are also expensive (approximately $200 each, so most people may not buy them for their personal use, but instead, will seek to share an office copy).

   It is good to read a reference guide from cover to cover if you have the time. The problem is not too many people have time to read the whole reference guide. Even if you do read the whole guide, you may not remember the important issues to pass the LEED exam. You need to reread the material several times before you can remember much of it.

   Reading the reference guide from cover to cover without a guidebook is a difficult and inefficient way of preparing for the LEED AP Exam, because you do NOT know what USGBC and GBCI are looking for in the exam.

2. The USGBC workshops and related handouts are concise, but they do not cover extra credits (exemplary performance). The workshops are expensive, costing approximately $450 each.

3. Various books published by a third party are available on Amazon, bn.com and books.google.com. However, most of them are not very helpful.

   There are many books on LEED, but not all are useful.

   LEED Exam Guides series will fill in the blanks and become a valuable, reliable source:

   a. They will give you more information for your money. Each of the books in the LEED Exam Guides series has more information than the related USGBC workshops.

   b. They are exam-oriented and more effective than the USGBC reference guides.

   c. They are better than most, if not all, of the other third-party books. They give you comprehensive study materials, sample questions and answers, mock exams and answers, and critical information on building LEED certification and going green. Other third-party books only give you a fraction of the information.

   d. They are comprehensive yet concise. They are small and easy to carry around. You can read them whenever you have a few extra minutes.

   e. They are great timesavers. I have highlighted the important information that you need to understand and MEMORIZE. I also make some acronyms and short sentences to help you easily remember the credit names.

It should take you about 1 or 2 weeks of full-time study to pass each of the LEED exams. I have met people who have spent 40 hours to study and passed the exams.

You can find sample texts and other information on the LEED Exam Guides series in customer discussion sections under each of my book's listing on Amazon, bn.com and books.google.com.

What others are saying about *LEED GA Exam Guide* (Book 2, LEED Exam Guide series):

"Finally! A comprehensive study tool for LEED GA Prep!

"I took the 1-day Green LEED GA course and walked away with a power point binder printed in very small print—which was missing MUCH of the required information (although I didn't know it at the time). I studied my little heart out and took the test, only to fail it by 1 point. Turns out I did NOT study all the material I needed to in order to pass the test. I found this book, read it, marked it up, retook the test, and passed it with a 95%. Look, we all know the LEED GA exam is new and the resources for study are VERY limited. This one is the VERY best out there right now. I highly recommend it."
—ConsultantVA

"Complete overview for the LEED GA exam

"I studied this book for about 3 days and passed the exam … if you are truly interested in learning about the LEED system and green building design, this is a great place to start."
—K.A. Evans

"A Wonderful Guide for the LEED GA Exam

"After deciding to take the LEED Green Associate exam, I started to look for the best possible study materials and resources. From what I thought would be a relatively easy task, it turned into a tedious endeavor. I realized that there are vast amounts of third-party guides and handbooks. Since the official sites offer little to no help, it became clear to me that my best chance to succeed and pass this exam would be to find the most comprehensive study guide that would not only teach me the topics, but would also give me a great background and understanding of what LEED actually is. Once I stumbled upon Mr. Chen's book, all my needs were answered. This is a great study guide that will give the reader the most complete view of the LEED exam and all that it entails.

"The book is written in an easy-to-understand language and brings up great examples, tying the material to the real world. The information is presented in a coherent and logical way, which optimizes the learning process and does not go into details that will not be needed for the LEED Green Associate Exam, as many other guides do. This book stays dead on topic and keeps the reader interested in the material.

"I highly recommend this book to anyone that is considering the LEED Green Associate Exam. I learned a great deal from this guide, and I am feeling very confident about my chances for passing my upcoming exam."
—Pavel Geystrin

"Easy to read, easy to understand

"I have read through the book once and found it to be the perfect study guide for me. The author does a great job of helping you get into the right frame of mind for the content of the exam. I had started by studying the Green Building Design and Construction reference guide for LEED projects produced by the USGBC. That was the wrong approach, simply too much information with very little retention. At 636 pages in textbook format, it would have been a daunting task to get through it. Gang Chen breaks down the points, helping to minimize the amount of information but maximizing the content I was able to absorb. I plan on going through the book a few more times, and I now believe I have the right information to pass the LEED Green Associate Exam."
—**Brian Hochstein**

"All in one—LEED GA prep material

"Since the LEED Green Associate exam is a newer addition by USGBC, there is not much information regarding study material for this exam. When I started looking around for material, I got really confused about what material I should buy. This LEED GA guide by Gang Chen is an answer to all my worries! It is a very precise book with lots of information, like how to approach the exam, what to study and what to skip, links to online material, and tips and tricks for passing the exam. It is like the 'one stop shop' for the LEED Green Associate Exam. I think this book can also be a good reference guide for green building professionals. A must-have!"
—**SwatiD**

"An ESSENTIAL LEED GA Exam Reference Guide

"This book is an invaluable tool in preparation for the LEED Green Associate (GA) Exam. As a practicing professional in the consulting realm, I found this book to be all-inclusive of the preparatory material needed for sitting the exam. The information provides clarity to the fundamental and advanced concepts of what LEED aims to achieve. A tremendous benefit is the connectivity of the concepts with real-world applications.

"The author, Gang Chen, provides a vast amount of knowledge in a very clear, concise, and logical media. For those that have not picked up a textbook in a while, it is very manageable to extract the needed information from this book. If you are taking the exam, do yourself a favor and purchase a copy of this great guide. Applicable fields: Civil Engineering, Architectural Design, MEP, and General Land Development."
—**Edwin L. Tamang**

Note: Other books in the **LEED Exam Guides series** are in the process of being produced. At least **one book will eventually be produced for each of the LEED exams.** The series include:

LEED v4 Green Associate Exam Guide (LEED GA): *Comprehensive Study Materials, Sample Questions, Mock Exam, Green Building LEED Certification, and Sustainability*, LEED Exam Guide series, ArchiteG.com. Latest Edition.

LEED GA MOCK EXAMS (LEED v4): *Questions, Answers, and Explanations: A Must-Have for the LEED Green Associate Exam, Green Building LEED Certification, and Sustainability,* LEED Exam Guide series, ArchiteG.com. Latest Edition

LEED v4 BD&C EXAM GUIDE: *A Must-Have for the LEED AP BD+C Exam: Comprehensive Study Materials, Sample Questions, Mock Exam, Green Building Design and Construction, LEED Certification, and Sustainability,* LEED Exam Guide series, ArchiteG.com. Latest Edition.

LEED v4 BD&C MOCK EXAMS: *Questions, Answers, and Explanations: A Must-Have for the LEED AP BD+C Exam, Green Building LEED Certification, and Sustainability,* LEED Exam Guide series, ArchiteG.com. Latest Edition.

LEED ID&C Exam Guide: *A Must-Have for the LEED AP ID+C Exam: Study Materials, Sample Questions, Green Interior Design and Construction, Green Building LEED Certification, and Sustainability,* LEED Exam Guide series, ArchiteG.com. Latest Edition.

LEED ID&C Mock Exam: *Questions, Answers, and Explanations: A Must-Have for the LEED AP ID+C Exam, Green Interior Design and Construction, Green Building LEED Certification, and Sustainability,* LEED Exam Guide series, ArchiteG.com. Latest Edition.

LEED O&M MOCK EXAMS: *Questions, Answers, and Explanations: A Must-Have for the LEED O&M Exam, Green Building LEED Certification, and Sustainability,* LEED Exam Guide series, ArchiteG.com. Latest Edition.

LEED O&M EXAM GUIDE: *A Must-Have for the LEED AP O+M Exam: Comprehensive Study Materials, Sample Questions, Mock Exam, Green Building Operations and Maintenance, LEED Certification, and Sustainability,* LEED Exam Guide series, ArchiteG.com. Latest Edition.

LEED HOMES EXAM GUIDE: *A Must-Have for the LEED AP Homes Exam: Comprehensive Study Materials, Sample Questions, Mock Exam, Green Building LEED Certification, and Sustainability,* LEED Exam Guide series, ArchiteG.com. Latest Edition.

LEED ND EXAM GUIDE: *A Must-Have for the LEED AP Neighborhood Development Exam: Comprehensive Study Materials, Sample Questions, Mock Exam, Green Building LEED Certification, and Sustainability,* LEED Exam Guide series, ArchiteG.com. Latest Edition.

How to order these books:
You can order the books listed above at:
http://www.GreenExamEducation.com

OR
http://www.ArchiteG.com

Building Construction

Project Management, Construction Administration, Drawings, Specs, Detailing Tips, Schedules, Checklists, and Secrets Others Don't Tell You (Architectural Practice Simplified, 2nd edition)

Learn the Tips, Become One of Those Who Know Building Construction and Architectural Practice, and Thrive!

For architectural practice and building design and construction industry, there are two kinds of people: those who know, and those who don't. The tips of building design and construction and project management have been undercover—until now.

Most of the existing books on building construction and architectural practice are too expensive, too complicated, and too long to be practical and helpful. This book simplifies the process to make it easier to understand and uncovers the tips of building design and construction and project management. It sets up a solid foundation and fundamental framework for this field. It covers every aspect of building construction and architectural practice in plain and concise language and introduces it to all people. Through practical case studies, it demonstrates the efficient and proper ways to handle various issues and problems in architectural practice and building design and construction industry.

It is for ordinary people and aspiring young architects as well as seasoned professionals in the construction industry. For ordinary people, it uncovers the tips of building construction; for aspiring architects, it works as a construction industry survival guide and a guidebook to shorten the process in mastering architectural practice and climbing up the professional ladder; for seasoned architects, it has many checklists to refresh their memory. It is an indispensable reference book for ordinary people, architectural students, interns, drafters, designers, seasoned architects, engineers, construction administrators, superintendents, construction managers, contractors, and developers.

You will learn:
1. How to develop your business and work with your client.
2. The entire process of building design and construction, including programming, entitlement, schematic design, design development, construction documents, bidding, and construction administration.
3. How to coordinate with governing agencies, including a county's health department and a city's planning, building, fire, public works departments, etc.
4. How to coordinate with your consultants, including soils, civil, structural, electrical, mechanical, plumbing engineers, landscape architects, etc.
5. How to create and use your own checklists to do quality control of your construction documents.
6. How to use various logs (i.e., RFI log, submittal log, field visit log, etc.) and lists (contact list, document control list, distribution list, etc.) to organize and simplify your work.
7. How to respond to RFI, issue CCDs, review change orders, submittals, etc.
8. How to make your architectural practice a profitable and successful business.

Planting Design Illustrated
A Must-Have for Landscape Architecture: A Holistic Garden Design Guide with Architectural and Horticultural Insight, and Ideas from Famous Gardens in Major Civilizations

One of the most significant books on landscaping!

This is one of the most comprehensive books on planting design. It fills in the blanks of the field and introduces poetry, painting, and symbolism into planting design. It covers in detail the two major systems of planting design: formal planting design and naturalistic planting design. It has numerous line drawings and photos to illustrate the planting design concepts and principles. Through in-depth discussions of historical precedents and practical case studies, it uncovers the fundamental design principles and concepts, as well as the underpinning philosophy for planting design. It is an indispensable reference book for landscape architecture students, designers, architects, urban planners, and ordinary garden lovers.

What Others Are Saying about *Planting Design Illustrated* …

"I found this book to be absolutely fascinating. You will need to concentrate while reading it, but the effort will be well worth your time."
—Bobbie Schwartz, former president of APLD (Association of Professional Landscape Designers) and author of *The Design Puzzle: Putting the Pieces Together*.

"This is a book that you have to read, and it is more than well worth your time. Gang Chen takes you well beyond what you will learn in other books about basic principles like color, texture, and mass."
—Jane Berger, editor & publisher of gardendesignonline

"As a longtime consumer of gardening books, I am impressed with Gang Chen's inclusion of new information on planting design theory for Chinese and Japanese gardens. Many gardening books discuss the beauty of Japanese gardens, and a few discuss the unique charms of Chinese gardens, but this one explains how Japanese and Chinese history, as well as geography and artistic traditions, bear on the development of each country's style. The material on traditional Western garden planting is thorough and inspiring, too. *Planting Design Illustrated* definitely rewards repeated reading and study. Any garden designer will read it with profit."
—Jan Whitner, editor of the *Washington Park Arboretum Bulletin*

"Enhanced with an annotated bibliography and informative appendices, *Planting Design Illustrated* offers an especially "reader friendly" and practical guide that makes it a very strongly recommended addition to personal, professional, academic, and community library gardening & landscaping reference collection and supplemental reading list."
—Midwest Book Review

"Where to start? *Planting Design Illustrated* is, above all, fascinating and refreshing! Not something the lay reader encounters every day, the book presents an unlikely topic in an easily digestible, easy-to-follow way. It is superbly organized with a comprehensive table of contents, bibliography, and appendices. The writing, though expertly informative, maintains its accessibility throughout and is a joy to read. The detailed and beautiful illustrations expanding on the concepts presented were my favorite portion. One of the finest books I've encountered in this contest in the past 5 years."

—Writer's Digest 16th Annual International Self-Published Book Awards Judge's Commentary

"The work in my view has incredible application to planting design generally and a system approach to what is a very difficult subject to teach, at least in my experience. Also featured is a very beautiful philosophy of garden design principles bordering poetry. It's my strong conviction that this work needs to see the light of day by being published for the use of professionals, students & garden enthusiasts."

—Donald C. Brinkerhoff, FASLA, chairman and CEO of Lifescapes International, Inc.

Index

Notes